Rancidity in Foods

*Dedicated in memory of Dr. Terry Galliard,
a valued colleague and scientist.*

Rancidity in Foods

Third edition

edited by

J. C. Allen
Chief Executive
Manchester Science Park
Manchester
UK

and

R. J. Hamilton
Director, Chemical & Physical Sciences
Liverpool John Moores University
Liverpool
UK

BLACKIE ACADEMIC & PROFESSIONAL
An Imprint of Chapman & Hall
London · Glasgow · Weinheim · New York · Tokyo · Melbourne · Madras

Published by
Blackie Academic and Professional, an imprint of Chapman & Hall,
Wester Cleddens Road, Bishopbriggs, Glasgow G64 2NZ

Chapman & Hall, 2-6 Boundary Row, London SE1 8HN, UK

Blackie Academic & Professional, Wester Cleddens Road, Bishopbriggs, Glasgow G64 2NZ, UK

Chapman & Hall GmbH, Pappelallee 3, 69469 Weinheim, Germany

Chapman & Hall USA, One Penn Plaza, 41st Floor, New York NY 10119, USA

Chapman & Hall Japan, ITP-Japan, Kyowa Building, 3F, 2-2-1 Hirakawa-cho, Chiyoda-ku, Tokyo 102, Japan

DA Book (Aust.) Pty Ltd, 648 Whitehorse Road, Mitcham 3132, Victoria, Australia

Chapman & Hall India, R. Seshadri, 32 Second Main Road, CIT East, Madras 600 035, India

First edition 1983
Second edition 1989
This edition 1994

© 1994 Chapman & Hall

Typeset in 10/12pt Times by Keytec Typesetting Ltd, Bridport, Dorset
Printed in Great Britain by St Edmundsbury Press, Bury St. Edmunds, Suffolk

ISBN 0 7514 0219 2

Apart from any fair dealing for the purposes of research or private study, or criticism or review, as permitted under the UK Copyright Designs and Patents Act, 1988, this publication may not be reproduced, stored, or transmitted, in any form or by any means, without the prior permission in writing of the publishers, or in the case of reprographic reproduction only in accordance with the terms of the licences issued by the Copyright Licensing Agency in the UK, or in accordance with the terms of licences issued by the appropriate Reproduction Rights Organization outside the UK. Enquiries concerning reproduction outside the terms stated here should be sent to the publishers at the Glasgow address printed on this page.

The publisher makes no representation, express or implied, with regard to the accuracy of the information contained in this book and cannot accept any legal responsibility or liability for any errors or omissions that may be made.

A catalogue record for this book is available from the British Library
Library of Congress Catalog Card Number: 94-71806

∞Printed on acid-free text paper, manufactured in accordance with ANSI/NISO Z39.48-1992 (Permanence of Paper)

Preface to the third edition

We were pleased when the publishers approached us for a third edition of this book, since it was clear that it had fulfilled its objective in providing a convenient and useful guide to the causes and prevention of rancidity in foods.

This edition has been largely rewritten and substantially enlarged. We have additional chapters on spectrophotometric and chromatographic analysis of oxidation, on rancidity in fish and a new chapter on rancidity in confectionery products. On the principle that prevention is better than cure, we have included a description of HACCP in the avoidance of rancidity. We have also added a chapter on legislative and labelling aspects. All the contributors are experts in their fields, and we are grateful to them for sharing our objective of producing a practical handbook rather than a theoretical treatise.

During the course of preparation we learnt of the serious illness of Dr. Terry Galliard, who was a keen supporter of our efforts from the beginning. His chapter on cereals has been a significant contribution to previous editions of this book. He was working on its revision right up to his death, and the news of it was a great sadness to us.

We are therefore honoured that Terry's family have kindly allowed us to dedicate this third edition to him, in memory of a good colleague and of his substantial and practical contribution to food science.

J. C. A.
R. J. H.

Contributors

J. C. Allen	Manchester Science Park Ltd, Enterprise House, Manchester Science Park, Lloyd Street North, Manchester, M15 6SE
K. G. Berger	17 Grosvenor Road, London, W4 4EQ
P. P. Coppen	May and Baker Ltd, Rainham Road South, Dagenham, Essex, RM10 7XS
A. C. Davies	Craigmillar Ltd., Craigmillar House, Stadium Road, Bromborough, Wirral, L62 3NT
A. Frampton	United Biscuits (UK) Ltd, The Chocolate Works, Bishopthorpe Road, York, YO1 1YE
T. Galliard	23 Old Windmill Way, Long Crandon, Aylesbury, Bucks, HP18 9BQ
M. H. Gordon	Department of Food Science and Technology, The University of Reading, Whiteknights, PO Box 226, Reading, RG6 2AP
R. J. Hamilton	Liverpool John Moores University, Byrom Street, Liverpool, L3 3AF
P. Harris	Unilever Research, Colworth House, Sharnbrook, Beds, MK44 1LQ
B. J. F. Hudson	Department of Food Science and Technology, The University of Reading, Whiteknights, PO Box 226, Reading, RG6 2AP
C. Humphries	CWS Secretariat, PO Box 53, New Century House, Manchester, M60 4ES
M. Livermore	Craigmillar Ltd, Craigmillar House, Stadium Road, Bromborough, Wirral, L62 3NT
J. Löliger	Nestle Ltd Research Centre, Nestec Ltd, PO Box 44, CH-1000, Lausanne 26, Switzerland
F. B. Padley	Unilever Research, Colworth House, Sharnbrook, Beds, MK44 1LQ

E. Prior	Nestlé Ltd Research Centre, Nestec Ltd, Vers-chez-les-Blanc, PO Box 44, CH-1000, Lausanne 26, Switzerland
M. D. Ranken	Michael Ranken Services, 9 Alexandra Road, Epsom, Surrey, KT17 4HB
J. B. Rossell	Leatherhead Food Research Association, Leatherhead, Surrey, KT22 7RY
T. A. B. Sanders	Department of Food and Nutritional Sciences, King's College, University of London, Campden Hill Road, London, W8 7AH
J. Tall	Unilever Research, Colworth House, Sharnbrook, Beds, MK44 1LQ

Contents

1 The chemistry of rancidity in foods   1
R. J. HAMILTON

   1.1 Introduction   1
   1.2 Rancidity   2
   1.3 Hydroperoxide formation   8
       1.3.1 Classical free radical route   8
       1.3.2 Induction period   11
       1.3.3 Photo-oxidation route   12
       1.3.4 Lipoxygenase route   14
   1.4 Secondary reaction products   14
       1.4.1 Aldehydes   15
       1.4.2 Alcohols   16
       1.4.3 Hydrocarbons   17
   1.5 Hydrolytic rancidity   17
   1.6 Antioxidants   18
   References   21
   Further reading   21

2 Measurement of rancidity   22
J. B. ROSSELL

   2.1 Introduction   22
   2.2 Measurement of hydrolytic rancidity   23
   2.3 Measurement of ketonic rancidity   26
   2.4 Measurement of oxidative rancidity   26
       2.4.1 Peroxide value (PV)   27
       2.4.2 Anisidine value (AV)   29
       2.4.3 Totox value   31
       2.4.4 Thiobarbituric acid (TBA) test   31
       2.4.5 Kreis test (rancidity index)   33
       2.4.6 Other chemical methods   33
       2.4.7 Physical methods   34
       2.4.8 Chromatographic methods   35
   2.5 Measurement of resistance to oxidative rancidity   37
       2.5.1 Schaal oven test   38
       2.5.2 Sylvester test   39
       2.5.3 FIRA–Astell apparatus   40
       2.5.4 The Oxidograph   43
       2.5.5 Swift test   45
       2.5.6 Rancimat apparatus   46
       2.5.7 OSI apparatus   49
   Acknowledgements   50
   References   51

3 Evaluation of oxidative rancidity   54
B. J. F. HUDSON and M. H. GORDON

   3.1 Introduction   54
   3.2 Sensory evaluation of rancidity   55

3.3	Estimation of oxidation artefacts		56
	3.3.1 Estimation of specific off-flavours		56
	3.3.2 Estimation of other end-products		56
3.4	Measurement of induction periods		57
3.5	Changes in iodine value or in component fatty acids		60
3.6	Peroxide values		60
3.7	Newer methods based on oxidation intermediates		61
3.8	Methods based on physical measurements		62
	3.8.1 Thermogravimetric analysis		62
	3.8.2 Chemiluminescence		62
	3.8.3 Other physical measurements		62
3.9	Correlations between methods for evaluating rancidity		63
3.10	Accelerated test strategy		64
3.11	Conclusions		65
References			65
Further reading			67

4 Practical measures to minimise rancidity in processing and storage
K. G. BERGER
68

4.1	Introduction		68
4.2	Design factors		69
	4.2.1 Storage tanks and process vessels		69
	4.2.2 Tank heating		71
	4.2.3 Pipelines		72
	4.2.4 Pumps		00
	4.2.5 Storage in drums		72
	4.2.6 Retail packaging		74
4.3	Operational factors		76
	4.3.1 Handling of bulk stock supplies		76
	4.3.2 Effect of dissolved oxygen		76
	4.3.3 Temperature control		81
Acknowledgements			82
References			82

5 The use of antioxidants
P. P. COPPEN
84

5.1	Basic principles and definitions		84
5.2	Why use an antioxidant?		85
5.3	Requirements of an ideal antioxidant		85
5.4	Oxidation of lard		86
5.5	Synergism		89
5.6	Popular misconceptions		90
5.7	Properties of the common antioxidants		90
	5.7.1 Tocopherol		91
	5.7.2 Ascorbyl palmitate		91
	5.7.3 BHA		94
	5.7.4 BHT		95
	5.7.5 Gallate esters		95
	5.7.6 TBHQ		96
5.8	Carry-through		97
5.9	Practical aspects of using antioxidants		99
5.10	Typical applications of antioxidants		100
5.11	Permitted rates of use		101
5.12	The future		101

| | Acknowledgement | 103 |
| | References | 103 |

6 Spectrophotometric and chromatographic assays — 104
E. PRIOR and J. LÖLIGER

6.1	Introduction	104
6.2	Substrate	105
6.3	Radicals	106
6.4	Hydroperoxides	106
	6.4.1 Spectrophotometric methods	106
	6.4.2 HPLC methods	108
	6.4.3 GC methods	114
	6.4.4 Indirect methods	115
	6.4.5 Other spectrometric methods	116
6.5	Secondary decomposition products	117
	6.5.1 Spectrophotometric methods	118
	6.5.2 HPLC methods	119
	6.5.3 GC methods	119
6.6	Fluorescent products	124
	References	126

7 Nutritional aspects of rancidity — 128
T. A. B. SANDERS

7.1	Introduction	128
7.2	Toxic components	128
7.3	Dietary sources of oxidised fats	129
7.4	Food processes that lead to the oxidation of fats	132
	7.4.1 Thermally oxidised fats	132
	7.4.2 Food irradiation	133
7.5	Biochemical effects resulting from lipid peroxides	134
7.6	Acute effects	134
7.7	Chronic effects	135
7.8	Toxicological studies on thermally oxidised fats	136
7.9	Possible long-term effects associated with the consumption of oxidised fats	137
7.10	Conclusion	139
	References	139

8 Rancidity in cereal products — 141
T. GALLIARD

8.1	Introduction	141
8.2	Hydrolytic rancidity	142
	8.2.1 Wheat (bran) lipase	143
	8.2.2 Lipase activity and hydrolytic rancidity in other cereal grains and products	146
8.3	Oxidative rancidity	147
	8.3.1 Enzymic oxidation (lipoxygenase action)	148
	8.3.2 Non-enzymic oxidation of cereal lipids	152
8.4	Factors affecting rancidity in cereal products	153
	8.4.1 Raw material quality	153
	8.4.2 Processing conditions	153
	8.4.3 Storage conditions	154
	8.4.4 Atmosphere	154
	8.4.5 Inhibitors	154
	8.4.6 Particle size	155
	8.4.7 Mixed products	155

	8.5	Conclusions	155
	References	156
	Further reading	156
	Appendix 1: measurement of wheat bran lipase activity in flour milling products	156
	Appendix 2: determination of degree of deterioration of flour milling products	157
	Appendix 3: assay for lipoxygenase activity in cereal extracts	158

9 Prevention of rancidity in confectionery and biscuits—a hazard analysis critical control point (HACCP) approach   160
A. FRAMPTON

9.1	Introduction	160
9.2	Ingredients	161
	9.2.1	Oils and fats	161
	9.2.2	Stabilised cereals (e.g. oats)	162
	9.2.3	Oleo resin extracts	163
9.3	Packaging materials	163
9.4	Equipment	164
9.5	Processing	164
	9.5.1	Commercial buying policy	164
	9.5.2	Reduced stock levels	164
	9.5.3	Reduction in blending at receiving factories	165
	9.5.4	Increased use of bulk deliveries	165
9.6	Shelf life	165
	9.6.1	Pursuit of freshness—'own label' policy	165
	9.6.2	Minimum durability—'best before'	166
9.7	HACCP	166
	9.7.1	Product acceptability	167
	9.7.2	Description of the hazards	167
	9.7.3	Identification of critical control points	168
	9.7.4	Critical control points	169
	9.7.5	Ingredients	170
	9.7.6	The process	172
	9.7.7	Finished product	172
	9.7.8	Packaging	173
	9.7.9	Distribution/point of sale	174
	9.7.10	Identification of control criteria and monitoring frequency	174
	9.7.11	Quantifying level of concern	175
	9.7.12	Agreement on procedures in the event of a deviation	175
	9.7.13	Verification	176
9.8	Conclusions	176
	References	177
	Further reading	177
	Appendix	177

10 Rancidity in dairy products   179
J. C. ALLEN

10.1	Introduction	179
10.2	Lipolytic rancidity	179
	10.2.1	Measurement of lipolysis	180
	10.2.2	Lipolysis in milk	180
	10.2.3	Lipolysis in dairy products	182
10.3	Oxidative rancidity	184
	10.3.1	Measurement of oxidative rancidity	185
	10.3.2	Oxidative rancidity in milk	185
	10.3.3	Oxidative rancidity in dairy products	187

	Acknowledgements	188
	References	189
	Further reading	189

11 Rancidity in meats
M. D. RANKEN
191

11.1	Introduction	191
11.2	Features special to meat	191
11.3	Types of rancidity	192
	11.3.1 Hydrolytic rancidity	192
	11.3.2 Oxidative rancidity	193
11.4	Special factors in meat which affect oxidative rancidity	194
	11.4.1 Meat pigments	194
	11.4.2 Effect of freezing	196
	11.4.3 Effect of salt	198
	11.4.4 Rancidity and nitrosamine formation	198
11.5	Rancidity and meat flavour	198
11.6	Control of rancidity	199
	11.6.1 Use of antioxidants	199
	11.6.2 Vacuum or modified atmosphere packaging	200
	11.6.3 Avoidance of pro-oxidants	201
	References	201

12 Legislation and labelling
C. HUMPHRIES
203

12.1	Introduction	203
12.2	Principles of food legislation	203
	12.2.1 Background	203
	12.2.2 Codex Alimentarius	204
	12.2.3 European Economic Community	205
12.3	Hygiene	205
	12.3.1 Food handling	205
	12.3.2 Cleaning and disinfection	206
	12.3.3 Hygiene regulation	206
	12.3.4 Quality systems	207
	12.3.5 HACCP	207
12.4	Irradiation	208
12.5	Packaging	208
12.6	Contaminants	209
12.7	Additives	209
	12.7.1 Safety assessment	210
	12.7.2 Relevant categories of additives	210
12.8	Labelling and presentation	217
	12.8.1 General labelling requirement	217
	12.8.2 Ingredients, durability and storage	217
	12.8.3 Claims	218
	References	220

13 Rancidity in creams and desserts
A. C. DAVIES and M. LIVERMORE
222

13.1	Introduction	222
13.2	Liquid creams	222
	13.2.1 Short life	222
	13.2.2 Long life	225

	13.3	Solid creams	226
	13.4	Powders	227
		13.4.1 Product	227
		13.4.2 Packaging	228
	13.5	Conclusions	228
	References		229

14 The control of rancidity in confectionery products   230
F. B. PADLEY

	14.1	Introduction	230
	14.2	The role of major ingredients	231
		14.2.1 Polysaccharides	231
		14.2.2 Protein	231
		14.2.3 Oils and fats	232
	14.3	The influence of water activity on flavour stability	237
	14.4	Individual confectionery fats	241
		14.4.1 Cocoa butter	241
		14.4.2 Butterfat and dairy ingredients	242
		14.4.3 Lauric fats, soapy and ketonic rancidity	246
		14.4.4 Confectionery fats other than CBE and lauric fats	248
	14.5	Nuts and seeds	248
	14.6	Other factors	254
	Acknowledgement		254
	References		254

15 Rancidity in fish   256
P. HARRIS and J. TALL

	15.1	Introduction	256
	15.2	Mechanisms	257
	15.3	Non-enzymic initiation	258
		15.3.1 Formation of active oxygen species	258
		15.3.2 Activated haem proteins	259
		15.3.3 Non-enzymic initiation in fish	260
	15.4	Enzyme initiation	260
	15.5	Microsomal enzyme lipid oxidation	260
	15.6	True enzyme initiation	261
	15.7	Measurements	265
		15.7.1 Chemical measurements	265
		15.7.2 Physical measurements	266
	15.8	Control of rancidity in fish	267
		15.8.1 Storage temperature	268
		15.8.2 Degree of butchery	268
		15.8.3 Oxygen control	268
	15.9	Conclusions	270
	References		270

Index   273

1 The chemistry of rancidity in foods
R. J. HAMILTON

1.1 Introduction

Although certain flavours can be obtained in foods via the sugars, or the proteins, it is normal to think of the off-flavours as being derived from the fats in foods. The composition of oils and fats can be given as 95% triglycerides with waxes, phosphoglycerides, sphingolipids, free fatty acids and hydrocarbons and non-lipids, e.g. vitamins, sterols, pigments and antioxidants making up the remainder. We shall see that the major (by far) component, the triglycerides (I), produces the flavour components.

$$\begin{array}{c} O \\ \| \\ R'C{-}OCH_2 \\ |O \\ |\| \\ CHOCR' \\ |O \\ |\| \\ CH_2OCR''' \end{array}$$

(I)

However, the flavours developed from phospholipids are believed to be different from triglycerides. These phosphoglycerides include lecithin (II), which has choline attached to the phosphatidic acid and cephalin, which is now recognised to be a mixture of phosphatidyl ethanolamine (III) and phosphatidyl serine (IV).

$$\begin{array}{ccc}
\text{CH}_2\text{OCOR} & \text{CH}_2\text{OCOR} & \text{CH}_2\text{OCOR} \\
| & | & | \\
\text{CHOCOR} & \text{CHOCOR} & \text{CHOCOR} \\
| \quad \text{O} & | \quad \text{O} & | \quad \text{O} \\
| \quad \| & | \quad \| & | \quad \| \\
\text{CH}_2\text{OP}\!-\!\text{OH} & \text{CH}_2\text{OP}\!-\!\text{OH} & \text{CH}_2\text{OP}\!-\!\text{OH} \\
| & | & | \\
\text{OCH}_2 & \text{OCH}_2 & \text{OCH}_2 \\
| & | & | \\
\text{CH}_2 & \text{CH}_2 & \text{CHCO}_2\text{H} \\
| & | & | \\
{}^+\text{N}(\text{CH}_3)_3 & \text{NH}_2 & \text{NH}_2 \\
\text{OH}^- & & \\
& & \\
\text{(II)} & \text{(III)} & \text{(IV)}
\end{array}$$

The physical properties and the fatty acid and triglyceride compositions of oils and fats are shown in Table 1.1–1.4. Only a selection of the more popular oils and fats is given since a fuller treatment is given in the literature suggested under Further reading.

The six major oils and fats on a world production basis are, for 1985, and estimated to be by 1995, as follows (in m metric tons, mt):

- Soyabean (1985) 13.9, (1995) 19.5;
- Palm (1985) 6.9, (1995) 13.9;
- Rape seed (1985) 6.0, (1995) 9.5;
- Sunflower (1985) 6.5, (1995) 8.9;
- Tallow (1985) 6.6, (1995) 7.3.

1.2 Rancidity

Rancidity can be considered to be based on the subjective organoleptic appraisal of the off-flavour quality of food. It is associated with a characteristic, unpalatable odour and flavour of the oils. The off-flavour can be caused by the absorption of the taints into the food or by contamination. The lipids act as reservoirs for the off-flavours. The rancidity can be caused by the changes that occur from reaction with atmospheric oxygen—the so-called oxidative rancidity. Finally the off-flavours can be produced by hydrolytic reactions which are catalysed by enzymes—the so-called hydrolytic rancidity. The hydrolytic reactions and the absorption effects can be minimised by cold storage, good transportation, careful packaging and sterilisation (see chapters 7–12), but oxidative rancidity, sometimes referred to as autoxidation, is not stopped by lowering the temperature of food storage. This is because autoxidation is

Table 1.1 Animal fats and oils

Fat	Source	Specific gravity 15°/15°	Refractive index n_D	Melting point (°C)	Saponification value	Iodine value	Acid value	Unsaponifiable (%)
A Solid fats								
Beef tallow	Cattle	0.937	$1.451^{60°}$	40–50	190–200	32–47	1–50	0.2–0.3
Butter	Cow's milk	0.940	$1.460^{25°}$	28–33	216–235	26–45	0.4–2.0	0.3–0.5
Ghee butter	Cow, buffalo or goat milk	0.863 $99°/15°$	$1.453^{60°}$	31–38	221–230	30–37	2.0–8.0	0.4
Lard	Pigs	0.936	$1.441^{60°}$	28–48	193–200	46–66	0.5–1.3	0.2–0.4
Mutton tallow	Sheep	0.950	$1.450^{60°}$	44–49	192–198	31–47	1.0–50.0	0.2–0.3
B Mammalian and marine oils								
Cod	*Gadus morrhua*	0.935			170–180	130–150	10–50	1–2
Cod liver	*Gadus morrhua*	0.925	$1.482^{15°}$		182–193	155–170	1–8	0.5–1.0
Herring	*Clupea harengus*	0.930	$1.472^{20°}$		183–190	123–146	2–50	1–4
Menhaden	*Alosa menhaden*	0.929	$1.481^{15°}$		189–195	160–180	4–12	0.6–1.6
Sardine	*Clupea sardinus*	0.933	$1.482^{18°}$		186–193	160–190	4–24	0.5–1.0

Adapted from Hamilton and Hamilton[1] (material originally published in Hilditch[2])

Table 1.2 Fatty acids present in animal fats and oils

Fatty acids present, in terms of percentage having number of carbon atoms indicated by (a).

Fat	10:0	12:0	14:0	16:0	16:1	18:0	18:1	18:2	18:3	20:0	20:1	22:0	22:1	Other fatty acids	S_3	S_2U SUS	S_2U SSU	SU_2 SUU	SU_2 USU	U_3	Type of analysis
A Animal fats																					
Beef tallow			3	26	6	17	43	4							12.6	30.6	13.1	31.9	3.4	8.4	L
Bone fat			3	32	3	15	43	3													
Butter	2	2	10	27	1	15	31	2						4:0 4%; 6:0 2%; 8:0 1%	19	58		20		3.0	C
Ghee butter	2	3	11	27		12	21	2						(a) 4:0 12%; 6:0 4%; 8:0 1%	28	54		15		3.0	C
Lard			2	26	4	14	43	10		1					4.9	1.9	27.0	11.2	38.9	16.1	L
Mutton tallow			4	21	3	31	36	5							16	46			31	6	L
B Animal oils																					
Cod			2	33	2	4	12	2	1		2			20:4 3%; 20:5 12%							
Cod liver			4	10	12	2	20				15		14	20:5 12%; 22:6 11%							
Herring			8	18	9	2	17	2			14		9	20:5 9%; 22:6 8%							
Menhaden			6	18	12	5	16	4	4					16 unsat. 5%; 18 unsat. 3%; 20:5 13%; 22:6 13%							
Sardine			6	10	13	2	24	2						(a) 20:3 26%; 22:3 19%				61		35	C

All fatty acid composition data are obtained by GLC except where indicated by (a). C, crystallisation. L, lipolysis. TLC, thin layer chromatography. S_3, triglycerides with three saturated fatty acids. S_2U, triglycerides with two saturated fatty acids and one unsaturated acid. (SUS), (SSU), positional isomers of S_2U. SU_2, triglycerides with one saturated acid and two unsaturated fatty acids. (USU), (SUU), positional isomers of SU_2. U_3, triglycerides with three unsaturated fatty acids. Adapted from Hamilton and Hamilton[1] (material originally published in Hilditch and Williams,[3] Gunstone[4] and Swern[5]).

Table 1.3 Vegetable fats and oils

A Solid fats

Fat	Botanical name of plant	Fat content (%)	Specific gravity (water at 15 °C, fat at temperature marked)	Refractive index[40] n_D	Melting point (°C)	Saponification value	Iodine value	Acid value	Unsaponified (%)
Cacao fat	Theobroma cacao	50–55 (beans)	0.970[15°]	1.456	32–35	190–198	33–40	1–4	0.2–1.0
Coconut oil	Cocos nucifera	63–65 (dried copra)	0.926[15°]	1.448	23–26	251–264	7–10	1–10	0.2–0.6
Mowrah (illipe) butter	Bassia latifolia (Madhuca latifolia)	50–55 (seeds)	0.920[15°]	1.460	23–31	188–200	53–70	5–50	1–3
Palm kernel oil	Elaeis guineensis	45–50 (kernel)	0.930[15°]	1.449	25–30	244–254	14–20	3–17	0.2–0.8
Palm oil	Elaeis guineensis	30–60 (fruit pulp)	0.924[15°]	1.457	38–45	196–202	48–56	2–15	0.2–0.5
Shea butter	Butyrospermum parkii	50 (kernels)	0.917[15°]	1.463	32–45	178–190	53–65	1–30	2–11

B Liquid oils

Oil	Botanical name of plant	Fat content (%)	Specific gravity[a] 15°/15°	Refractive index[b40] n_D	Saponification value	Iodine value	Hydroxyl value	Acid value	Unsaponified (%)
Blackcurrant	Ribes	35–40		1.479–1.481	185–195	173–182	–		0.1–1.0
Cottonseed oil	Gossypium spp.	22–28 (seeds)	0.923	1.470[25°]	191–196	100–112		0.2–4	0.6–0.8
Evening primrose	Oenothera biennis	20 (seeds)		1.479[1°]	192.8	155			1.5–2.0
Grapeseed	Vitis vinifera	40–50 (kernels)	0.923–0.926[20°/20°]	1.473–1.477	188–194	130–138			0–2.0
Groundnut oil	Arachis hypogaea	50 (germs)	0.918	1.462	189–196	84–102	3–9	0.1–6	0.2–0.8
Maize oil	Zea mays	31–33 (seeds)	0.924	1.466[40°]	187–196	117–130	–	2–6	0.8–2.9
Mustard (black) oil	Brassica nigra	25–36 (seeds)	0.920	1.469[25°]	173–180	101–112	–	1–7	0.7–1.5
Mustard (white) oil	Brassica alba	30–60 (fruit)	0.918	1.469[25°]	170–180	94–105	–	0.5–8	0.7–1.5
Olive oil	Olea europea sativa	32–45 (seeds)	0.916	1.462	187–196	80–90	5–11	0.2–6	0.5–1.3
Rape oil (Colza)	Brassica campestris		0.916	1.465	173–181	105–120	1–6	0.5–10	0.6–1.2
Rapeseed (Lear)	Brassica campestris	25–32		1.465–1.467	188–193	110–126			0.2–1.8
Safflower oil	Carthamus tinctorius	15–20	0.926	1.469[40°]	187–194	130–150	1–12	0.4–10	0.3–1.3
Soyabean oil	Soja japonica	22–32	0.926	1.468[40°]	189–195	124–133	–	0.2–0.6	0.3–3
Sunflower oil	Helianthus annus		0.924	1.466[40°]	186–194	127–136	–	0.6–4	0.3–0.5
High oleic sunflower	Helianthus annus								
Walnut oil	Juglans regia	63–65 (kernels)	0.925	1.475[25°]	190–197	140–150	–	4–10	0.2–0.4
Wheat oil	Triticum aestivum	12–17 (germs)	0.926	1.472[40°]	179–190	115–129	–	2–6	1–11

[a] At 15 °C (unless otherwise indicated), referred to water at 15 °C. [b] At 40 °C (unless otherwise indicated), referred to D sodium line. Adapted from Hamilton and Hamilton[1] (material originally published in Hilditch[2]).

Table 1.4 Fatty acids present in vegetable fats and oils

Fat/Oil	Fatty acids present, in terms of percentage having number of carbon atoms indicated											Other fatty acids	S_3	Triglyceride composition				Type of analysis
	10:0	12:0	14:0	16:0	16:1	18:0	18:1	18:2	18:3	20:0	20:1	22:0	22:1			S_2U SUS	SSU	SU_2 SUU USU	U_3	

	10:0	12:0	14:0	16:0	16:1	18:0	18:1	18:2	18:3	20:0	20:1	22:0	22:1	Other fatty acids	S_3	SUS	SSU	SUU	USU	U_3	Type of analysis
A Fats																					
Cacao				27		33	35	3		2					2.7	80.8	0.5	15.3		0.7	L
Coconut	7	47	17	8		4	5	2						8:0 10%	79.1	10.7	8.7	1.2	0.3	0	L
Mowrah (illipe) butter				23.7	0.2	24.1	37.6	14.4							3.5	80.4	0.7	14.6		0.7	L
Palm kernel	4	51	17	8		2	13	2						10:0 4%; 8:0 3%							
Palm oil		1		48		4	38	9							8.2	34.8	8.7	36.5	2.2	9.6	L
Shea butter				8		36	50	5						(a)	1.6	42.2	1.5	43.2	0.4	11.1	L
B Oils																					
Blackcurrant seed				6.0		1	10	48	13					18:3(n-6)8%							
Cottonseed			1.1	27.3	1.4	3.1	16.7	50.4							0.3	15.5	1.0	47.0	0.7	35.5	L
Evening primrose				7.0	tr	2.0	9.0	72.0	2.0ᵃ					18:3(n-6)16% 18:4 3%							
Grapeseed				6		1.0	16.0	71													
Groundnut				9.8	0.4	3.7	60.9	18.1						24:0 1.0%	0.1	9.3	0.6	8.8	46.9	47.5	L
Maize				12.6	0.8	1.8	30.0	54.3	0.5						0		5.0	33.0		62.0	L
Mustard (black) oil				3			8	14	18	1	7		43								
Mustard (white) oil				3	0.4	1	16	10	14		9		47								
Olive				14.0	2.0	2.0	64.0	16.0							0.1	4.8	0.2	34.3	0.1	60.5	L
Rape				3		2	22	15	14	1	15		28		0					82	C
Rapeseed (Lear)				4		1	51	30	12							18					
Safflower				6.6	0.6	3.4	12.2	77.0	0.2						0		2	26		72	L
Soyabean				12		3.6	23.7	51.4	8.8						0		7	34		59	L
Sunflower				6		3	27	64							0		3	25		72	L
High oleic sunflower				6		2	80	12													
Walnut				8		3	15	61	12												
Wheat				4		4	27	59	7												

For key see Table 1.2. Adapted from Hamilton and Hamilton[1] (material originally published in Hilditch and Williams,[3] Gunstone[4] and Swern[5]).

a chemical reaction with a low activation energy, 4–5 kcal mol^{-1}, for the first step and 6–14 kcal mol^{-1} for the second step.

Research into these problems concerning flavour deterioration has been pursued for many years and it has been given a boost by the recognition that such oxidations can cause damage to membranes, enzymes, vitamins and proteins, and that they may be involved in the ageing process.

Crushing or macerating tissue, either animal or vegetable, starts off lipolysis because lipolytic enzymes, lipases, are released, which act on lipids to release the free fatty acids as follows:

$$\begin{array}{c} CH_2OCOR' \\ | \\ CHOCOR'' \\ | \\ CH_2OCOR''' \end{array} + 3H_2O \xrightarrow{lipase} \begin{array}{c} CH_2OH \\ | \\ CHOH \\ | \\ CH_2OH \end{array} + R'CO_2H + R''CO_2H + R'''CO_2H$$

The free fatty acids and the triglycerides are capable of being oxidised by autoxidation or by enzymes called lipoxygenases. The oxidation is concerned primarily with the unsaturated fatty acids, of which oleic (V), linoleic (VI) and linolenic (VII) acids are the most abundant.

$$CH_3(CH_2)_7CH\overset{9}{-}CH(CH_2)_7\overset{1}{C}O_2H$$

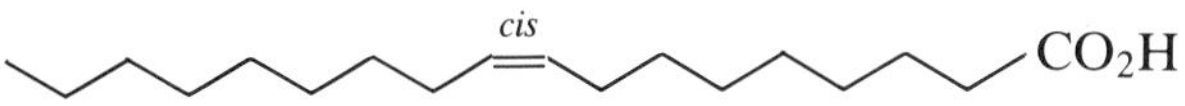

(V)

$$CH_3(CH_2)_4CH\overset{12}{=}CHCH_2CH\overset{9}{=}CH(CH_2)_7\overset{1}{C}O_2H$$

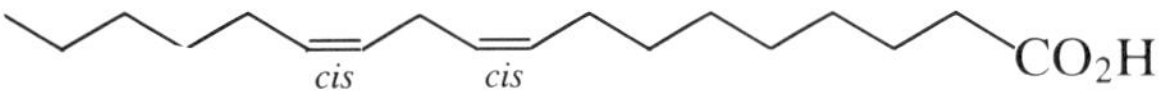

(VI)

$$CH_3CH_2CH\overset{15}{=}CHCH_2CH\overset{12}{=}CHCH_2CH\overset{9}{=}CH(CH_2)_7\overset{1}{C}O_2H$$

(VII)

It is accepted that the first product of oxidation is an intermediate, which is itself odourless, but which breaks down to smaller molecules which do produce the off-flavour. In autoxidation it is also known that the three unsaturated acids are oxidised at different rates, viz. linoleic acid is oxidised 64 times faster than oleic acid, and linolenic acid is oxidised 100 times faster than oleic acid. The production of the intermediate can occur by one of two mechanisms: the classical free radical mechanism, which can operate in the dark, or a photo-oxidation mechanism, which is initiated by exposure to light. The intermediate in both cases is a lipid hydroperoxide.

1.3 Hydroperoxide formation

1.3.1 Classical free radical route

The classical route depends on the production of free radicals R$^{\cdot}$ from lipid molecules RH by their interaction with oxygen in the presence of a catalyst, e.g. the initiation can occur by the action of external energy sources such as heat, light or high energy radiation or by chemical initiation involving metal ions or metalloproteins such as haem. The mechanism of these initiation steps is still not fully understood. The free radical R$^{\cdot}$ produced in the initiation steps can then react to form a lipid peroxy radical ROO$^{\cdot}$ which can react further to give the hydroperoxide ROOH. The second reaction of the propagation steps also provides a further free radical R$^{\cdot}$, making it a self-propagating chain process. In this way a small amount of catalyst, e.g. copper ions, can initiate the reaction, which then produces many hydroperoxide molecules, which ultimately break down to cause rancidity. The self-propagating chain can be stopped by termination reactions, where two radicals combine to give products which do not feed the propagating reactions.

Initiation: $\quad RH + O_2 \xrightarrow{catalyst} R^{\cdot} + {}^{\cdot}OOH$
$\quad\quad\quad\quad\quad\quad RH \xrightarrow{catalyst} R^{\cdot} + {}^{\cdot}H$
Propagation: $\quad R^{\cdot} + O_2 \rightarrow RO_2^{\cdot}$
$\quad\quad\quad\quad\quad\quad RO_2^{\cdot} + RH \rightarrow RO_2H + R^{\cdot}$
Termination: $\quad R^{\cdot} + R^{\cdot} \rightarrow R-R$
$\quad\quad\quad\quad\quad\quad RO_2^{\cdot} + R^{\cdot} \rightarrow RO_2R$

where RH = unsaturated lipid, R$^{\cdot}$ = lipid radical and RO$_2^{\cdot}$ = lipid peroxy radical.

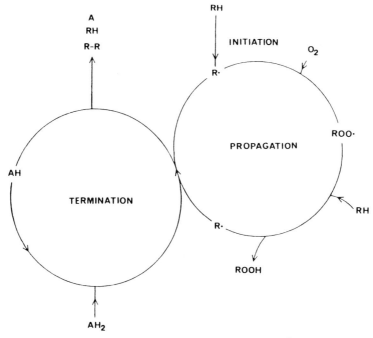

Figure 1.1 The classical free radical route.[6]

These reactions can be shown in diagrammatic form which stresses the cyclic nature of the process (Figure 1.1).

When this type of mechanism is applied to the autoxidation of methyl oleate (Figure 1.2), it can be seen that hydrogen abstraction on C-8 and C-11 forms two allylic radicals, each of which can be represented by two canonical forms. Such canonical forms allows explanation why not only the 8-hydroperoxide is obtained but also why the 10-hydroperoxide from one allylic radical and the 9- and 11-hydroperoxides from the other allylic radical are obtained. It is also our justification for not expecting hydrogen abstraction at any other position in the fatty acid chain. The double bond position is scrambled, i.e. there are other hydroperoxides present in addition to the Δ^9 hydroperoxide and the configuration may be changed from *cis* to *trans*. The autoxidation of methyl linoleate starts with the abstraction of a hydrogen at the doubly reactive methylene at C-11 (Figure 1.3). Again the hydroperoxide is found to be a mixture of conjugated dienes with the hydroperoxide group at positions 9 and 13. Methyl linolenate gives a hydroperoxide mixture with the hydroperoxide group at positions 9, 12, 13 and 16. The presence of such mixtures can be explained by assuming that there are canonical forms as indicated. Frankel[7] has suggested that the classical hydroperoxide theory (as shown

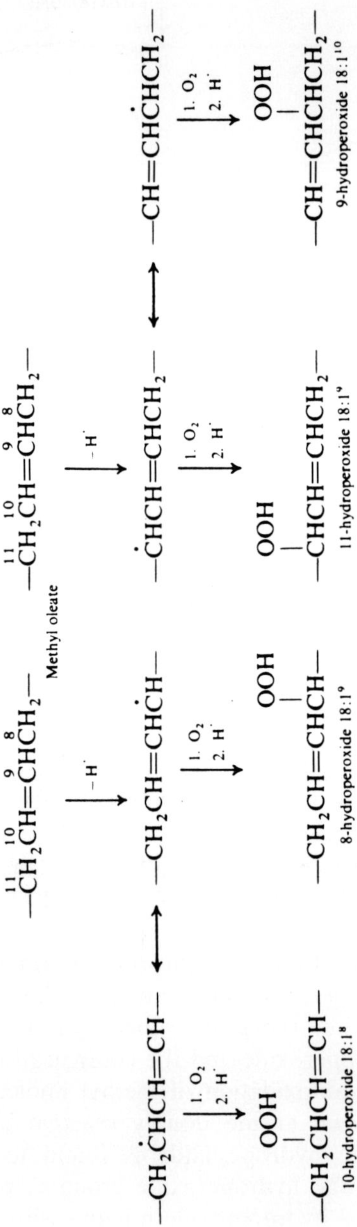

Figure 1.2 Autoxidation of methyl oleate.

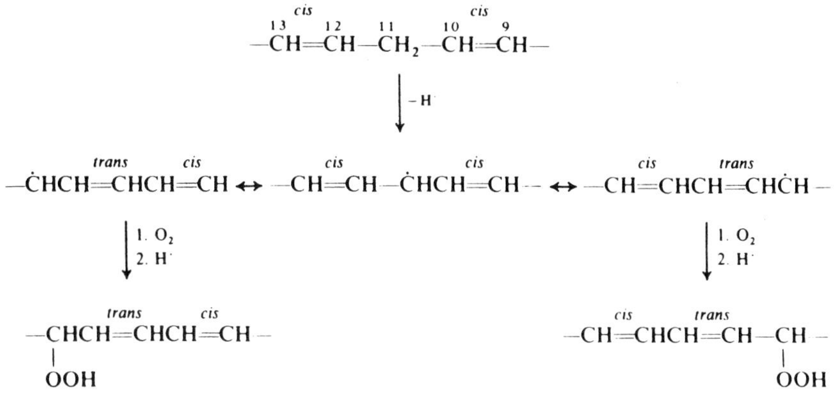

$$-\overset{\overset{\displaystyle cis}{}}{\underset{13}{CH}}=\underset{12}{CH}-\underset{11}{CH_2}-\overset{\overset{\displaystyle cis}{}}{\underset{10}{CH}}=\underset{9}{CH}-$$

$$\downarrow -H^{\cdot}$$

$$-\overset{trans}{\dot{C}H}\overset{cis}{CH}=CHCH=CH \leftrightarrow -CH\overset{cis}{=}CH-\overset{cis}{\dot{C}HCH}=CH- \leftrightarrow -CH\overset{cis}{=}CHCH\overset{trans}{=}CH\dot{C}H-$$

$$\downarrow \begin{matrix} 1.\ O_2 \\ 2.\ H^{\cdot} \end{matrix} \qquad\qquad\qquad\qquad\qquad\qquad\qquad \downarrow \begin{matrix} 1.\ O_2 \\ 2.\ H^{\cdot} \end{matrix}$$

$$\underset{\underset{\displaystyle OOH}{|}}{-\overset{trans}{C}H}\overset{cis}{CH}=CHCH=CH- \qquad\qquad\qquad -CH\overset{cis}{=}CHCH\overset{trans}{=}CH-\underset{\underset{\displaystyle OOH}{|}}{CH}-$$

Figure 1.3 Autoxidation of methyl linoleate.

in Figures 1.2 and 1.3) does not explain his results with regard to percentage composition of the hydroperoxide positional isomers.

1.3.2 Induction period

When the autoxidation of a fat is followed experimentally, e.g. by measuring the amount of oxygen absorbed or the peroxide value, it is found that the course of the oxidation shows two distinct phases. During the first phase, the oxidation goes slowly and at a uniform rate. After the oxidation has proceeded to a certain point the reaction enters a second phase, which has a rapidly accelerating rate of oxidation, and the eventual rate is many times greater than that observed in the initial phase. The oil starts to taste at the start of the second phase though expert tasters can sense the subtle changes in oil quality before this second phase is reached.

The initial phase is called the induction period. Figure 1.4 represents a typical induction period for corn oil which can be extended by performing

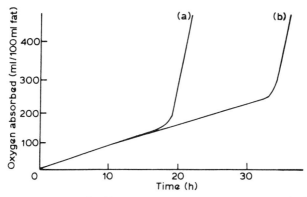

Figure 1.4 Oxygen uptake (a) in corn oil, and (b) in corn oil plus antioxidant.

the test a second time with an antioxidant present. With some oils the rapid change in the curve is not so clear cut as in Figure 1.4, i.e. the change in slope is much more gradual.

It is found that methyl linoleate reacts more quickly than methyl oleate and has a shorter induction period.

1.3.3 Photo-oxidation route

Photo-oxidation has now been recognised as an alternative to the free radical mechanism, because it is found that different hydroperoxides are formed when light and certain photosensitiser molecules are present. There had always been a worry over implicating oxygen, because of the concept of conservation-of-spin between reactant molecules. It was realised that singlet oxygen reacts 1500 times faster with methyl linoleate than does triplet oxygen, and that singlet oxygen reacts directly with double bonds by addition at either end of the double bond, producing an allylic hydroperoxide in which the double bond has been shifted (Figure 1.5).

This reaction is known as the 'ene' reaction. It can be shown that singlet oxygen has a relative reactivity in the following order: cholesterol

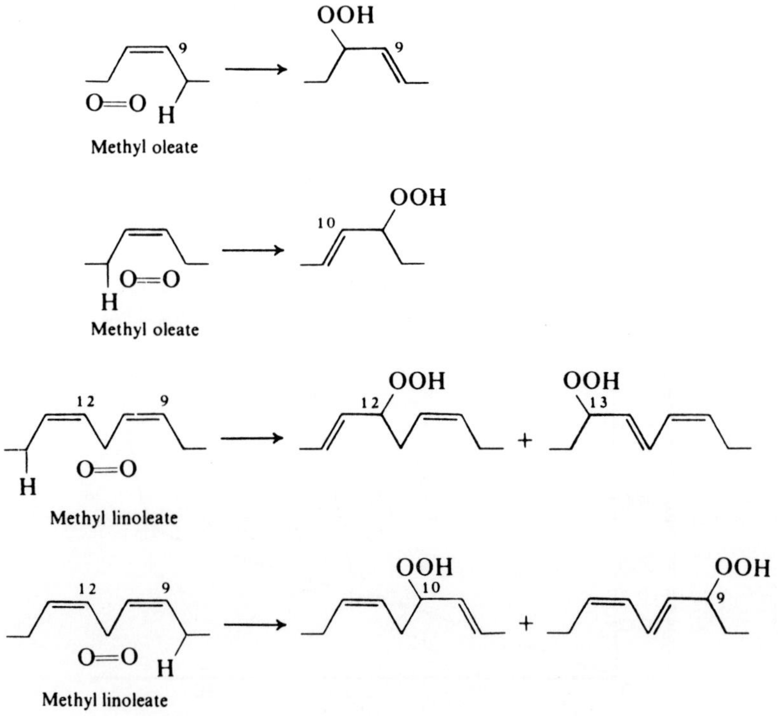

Figure 1.5 Photo-oxidation route of hydroperoxide formation.

(1.0), oleic acid (1.1), linoleic acid (1.9), linolenic acid (2.9) and arachidonic acid (3.5). Singlet oxygen is produced by light in the presence of sensitisers, e.g. chlorophyll, myoglobin, erythrosine, riboflavine and heavy metal ions. With these oxidations, no induction period is known. Two mechanisms have been mentioned for these photo-oxidations: Type I and Type II photo-oxidations.

Type I

$$^1Sens \xrightarrow{h\nu} {}^1Sens^* \rightarrow {}^3Sens^*$$

$$^3Sens^* + X \text{ (acceptor)} \rightarrow [\text{Intermediate I}]$$

$$[\text{Intermediate I}] + {}^3O_2 \rightarrow {}^1Sens + XO_2$$

Type I photo-oxidations lead to similar hydroperoxides to those found in the classical free radical route. In the presence of visible or ultraviolet radiation, the ground state sensitiser absorbs the energy and is converted to an excited singlet state ($^1Sens^*$). This excited singlet sensitiser emits fluorescent light and returns to ground state. An alternative route is by intersystem crossing (ISC) by which an excited triplet state of the sensitizer is formed (Figure 1.6).

This excited triplet state of the sensitiser molecule in Type I reacts with acceptor substrates to produce Intermediate I which may be a radical or a radical ion. This Intermediate I then reacts with triplet oxygen to form oxidised products. However, Type II photo-oxidations lead to quite different hydroperoxides and thus to different volatiles when they break down to give secondary oxidation products. In the Type II mechanism the excited triplet state of the sensitiser molecule reacts with triplet oxygen to form singlet oxygen.

Type II

$$^3Sens^* + {}^3O_2 \rightarrow {}^1O_2^* + {}^1Sens$$

$$^1O_2^* + RH \rightarrow ROOH$$

Type II oxidation can be inhibited by β-carotene and α-tocopherol

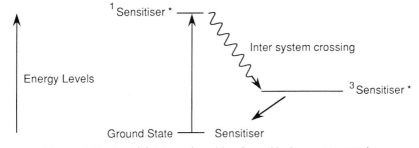

Figure 1.6 Excited triplet state of sensitiser formed by intersystem crossing.

which act as quenchers either by deactivating singlet oxygen to ground state oxygen or reacting with it. Some molecules can prevent Type II oxidation by reacting faster with singlet oxygen than the lipid molecule can react with singlet oxygen.

1.3.4 Lipoxygenase route

The enzyme lipoxygenase is now believed to be widely distributed throughout the plant and animal kingdoms. The basic stoichiometry of the lipoxygenase oxidation reaction is the same as for autoxidation, but in common with many enzyme reactions, lipoxygenase is very specific about the substrate and how the substrate is oxidised. Linoleic acid is oxidised at positions 9 and 13 by lipoxygenase isolated from most natural sources.

Lipoxygenase prefers free fatty acids as substrates, though there are lipoxygenase isoenzymes which can react with triglycerides. The regio-specificity of the reaction, i.e. the preference for certain positions on the carbon chain, and the stereospecificity, i.e. the preference for one of the two possible hydrogens at any specified carbon atom, are indicated in Figure 1.7. Further examples of this enzymic oxidation are found in chapter 7.

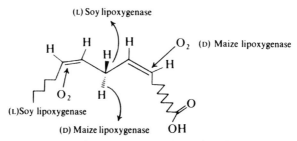

Figure 1.7 Stereospecific oxygenation of linoleic acid by lipoxygenase.

It is worthwhile repeating that each of these three routes, viz. classical autoxidation, photo-oxidation and lipoxygenase oxidation, leads to hydroperoxides which are relatively involatile and which do not produce an off-flavour. The conversion of the hydroperoxides into the final flavour components is less well understood than the production of hydroperoxides.

1.4 Secondary reaction products

Lipid hydroperoxides are very unstable and break down to an alkoxy free radical which decomposes, mainly by cleavage on either side of the carbon atom bearing the oxygen atom.

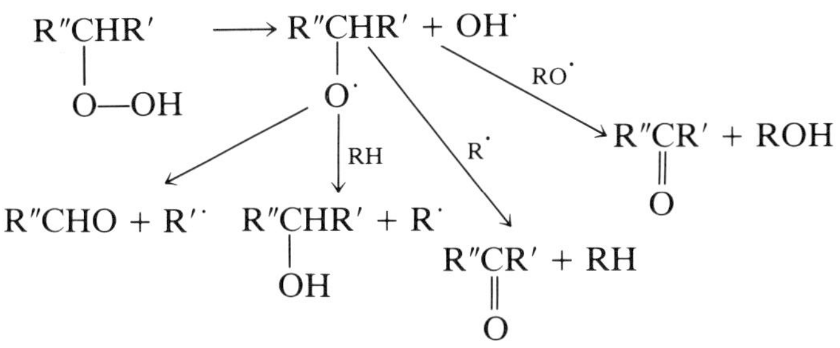

$$R''CHR' \longrightarrow R''CHR' + OH^{\cdot}$$
$$\underset{O-OH}{|} \quad \underset{O^{\cdot}}{|}$$

Scheme showing alkoxy radical pathways:

- R"CHR'–O–OH → R"CHR'(O·) + OH·
- R"CHR'(O·) + RH → R"CHO + R"·
- R"CHR'(O·) + RH → R"CHR'(OH) + R·
- R"CHR'(O·) + R· → R"CR'(=O) + RH
- R"CHR'(O·) + RO· → R"CR'(=O) + ROH

1.4.1 Aldehydes

The mechanism for the cleavage of the alkoxy free radical depends on the cleavage on either side of the carbon atom containing the oxygen atom. The two odd electrons produced on neigbouring atoms can then form the carbonyl double bond. The following example shows cleavage of 11-hydroperoxyoleic acid methyl ester:

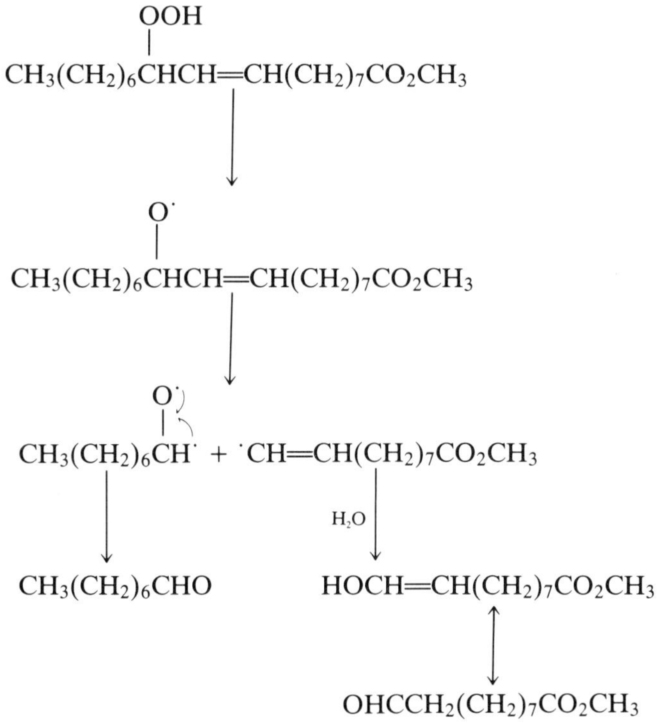

$$\underset{|}{OOH}$$
$$CH_3(CH_2)_6CHCH=CH(CH_2)_7CO_2CH_3$$

↓

$$\underset{|}{O^{\cdot}}$$
$$CH_3(CH_2)_6CHCH=CH(CH_2)_7CO_2CH_3$$

↓

$$\underset{|}{O^{\cdot}}$$
$$CH_3(CH_2)_6CH^{\cdot} + {}^{\cdot}CH=CH(CH_2)_7CO_2CH_3$$

↓ H_2O ↓

$CH_3(CH_2)_6CHO$ $HOCH=CH(CH_2)_7CO_2CH_3$

↑↓

$OHCCH_2(CH_2)_7CO_2CH_3$

Thus aldehyde groups can be produced directly and indirectly via an enol, which is simply the tautomeric form of an aldehyde (in this case giving octanal and methyl 10-oxodecanoate respectively). Clearly with the range of hydroperoxides available there are a great many aldehydes which can be produced.

Table 1.5 Flavours of aliphatic aldehydes

No. of carbon atoms	Homologous series		
	Saturated	2-Enals	2,4-Dienals
C_2	fresh, pungent		
C_3	fresh, milky		
C_4	↓	sweet, pungent	
C_5		sweet, green	
C_6	fresh, green	↓	
C_7	↓		sweet, oily
C_8	fresh, citrus	↓	
C_9		sweet, fatty, green	
C_{10}		↓	
C_{11}	fatty	sweet, fatty	
C_{12}	↓	↓	↓

Aldehydes give rise to flavours which are described as ranging from sweet and pungent to oxidised milk. The saturated aldehydes are said to contribute power, warmth, resonance, depth, roundness and freshness to the flavour, whilst the 2-enals and 2,4-dienals are said to contribute sweet, fruity or fatty and oily characters to the flavour. The saturated aldehydes are described as C_2 (fresh, pungent), C_3 (fresh, milky), through C_6 (fresh, green), C_8 (fresh, citrus) to C_{11} (fatty). Descriptions of the flavour contributions of saturated and unsaturated aldehydes are shown in Table 1.5.

1.4.2 Alcohols

The alcohols can be formed by a mechanism which is similar to that for aldehydes. The alkoxy free radical cleaves to give an aldehyde and a hydrocarbon free radical which can pick up an OH˙ radical to give the alcohol or alternatively pick up an H˙ radical to form a hydrocarbon:

$$CH_3(CH_2)_7 \overset{\overset{O˙}{|}}{C}HCH=CH(CH_2)_6CO_2CH_3$$

$$\downarrow$$

$$CH_3(CH_2)_7˙ + \overset{\overset{O}{\|}}{C}HCH=CH(CH_2)_6CO_2CH_3$$

$$\swarrow_{H˙} \qquad \searrow_{˙OH}$$

$$CH_2(CH_2)_6CH_3 \qquad CH_3(CH_2)_6CH_2OH$$

The alcohols are believed to contribute to the flavour in the same manner as the aldehydes, but in a milder way, ranging from the C_3 saturated alcohol which is described as solventy, non-descript, to C_6, described as grassy, green, to C_9, described as fatty, green.

1.4.3 Hydrocarbons

It is possible to postulate mechanisms similar to those for aldehydes and alcohols to account for the hydrocarbons. In addition, if we postulate that the H· radical is acquired from R'H, we form a new radical R''· with the result that further chain reactions can occur:

$$CH_3CH_2CH=CHCH_2\overset{\overset{\displaystyle O\cdot}{|}}{C}HCH=CHCH=CH(CH_2)_7CO_2CH_3$$

$$\downarrow$$

$$CH_3CH_2CH=CHCH_2^{\cdot} + \overset{\overset{\displaystyle O}{\|}}{C}HCH=CHCH=CH(CH_2)_7CO_2CH_3$$

$$\swarrow^{R'H} \qquad \searrow^{\cdot OH}$$

$$CH_3CH_2CH=CHCH_3 + R''\cdot \qquad CH_3CH_2CH=CHCH_2OH$$

When these general cleavage methods are applied to methyl linolenate hydroperoxides, a variety of products are obtained, some of which are shown in Table 1.6.

1.5 Hydrolytic rancidity

The three groups of molecules—aldehydes, alcohols and hydrocarbons—can be seen to arise from the hydroperoxides produced by photooxidation or by the classical autoxidation mechanism. Methyl ketones, the lactones and the esters may be formed primarily by hydrolytic reactions. Thus the glyceride molecule, under the action of heat and moisture, may break down to keto acids, which lose carbon dioxide readily:

$$\begin{array}{l} CH_2OCOR' \\ | \\ CHOCOR \\ | \\ CH_2OCOR \end{array} \xrightarrow{H_2O, \triangle} HO_2CCH_2\overset{\overset{\displaystyle O}{\|}}{C}R' \rightarrow CO_2 + CH_3\overset{\overset{\displaystyle O}{\|}}{C}R' \rightarrow CH_3\overset{\overset{\displaystyle OH}{|}}{C}HR'$$

The release of hydroxy fatty acids can provide the precursors for γ- or δ-lactones

$$\begin{array}{l} CH_2OCO(CH_2)_{16}CH_3 \\ | \\ CHOCO(CH_2)_{16}CH_3 \\ | \\ CH_2OCO(CH_2)_{16}CH_3 \end{array} \xrightarrow{\text{1. oxidation}}_{\text{2. H}_2\text{O/lipase}}$$

$$HO_2C(CH_2)_2CHOH(CH_2)_{13}CH_3 + HO_2C(CH_2)_3CHOH(CH_2)_{12}CH_3 +$$

$$\downarrow \qquad\qquad \downarrow \qquad\qquad \begin{array}{l} CH_2OH \\ | \\ CHOH \\ | \\ CH_2OH \end{array}$$

$$\underset{O}{\underset{\diagdown}{O=C}}\overset{(CH_2)_2}{\overset{\diagup\diagdown}{}}\underset{\diagup}{CH(CH_2)_{13}CH_3} + \underset{O}{\underset{\diagdown}{O=C}}\overset{(CH_2)_3}{\overset{\diagup\diagdown}{}}\underset{\diagup}{CH(CH_2)_{12}CH_3}$$

It is also believed that the hydrolytic reactions, including lipolysis, provide the free oleic, linoleic or linolenic acids which can undergo more rapid autoxidation.

Methyl ketones contribute a piercing sweet fruitiness, ranging from C_3, pungent, sweet, through C_7 blue cheesy to C_{11} fatty, sweet. The aliphatic acids contribute to the flavour by being sour, fruity, cheesy or animal-like. Their contribution ranges from C_2 vinegary, C_3 sour, Swiss cheesy, C_4 sweaty cheesy, C_8 goat cheesy, C_9 paraffinic to C_{14}–C_{18} with very little odour. The contributions of each of these components to the flavour of a food depends on the flavour threshold. Clearly a compound, e.g. 1-octen-3-ol which has a value of 0.001 (see table 1.7) in water will create a very big impact on the flavour of a food containing even a very small amount.

There are suggestions[8] that methyl ketones can be produced, provided water is present, by micro-organisms. Kinderlerer has noted that pentan-2-one at the trace level in rancid coconut oil produces a 'pear drops' flavour note. Hexan-2-one, by contrast, at the trace level, is quoted as 'ethereal' whilst heptan-2-one at 5.1 mg g^{-1} of oil is considered as 'rancid almonds'. Nonan-2-one at 3.0 mg g^{-1} of oil and undecan-2-one at 1.3 mg g^{-1} of oil are said to be 'weakly turpentine'.

1.6 Antioxidants

Autoxidation can be inhibited or retarded by adding a low concentration of an antioxidant (AH) which can interfere with either chain propagation

Table 1.6 Decomposition of methyl linolenate hydroperoxides

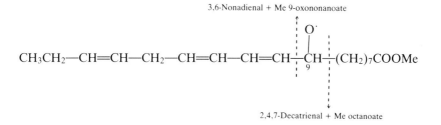

Cleavage at the arrow yields the products indicated.

or initiation:

$$ROO^\cdot + AH \rightarrow ROOH + A^\cdot$$

$$\left.\begin{array}{r}A^\cdot + ROO^\cdot \rightarrow \\ A^\cdot + A^\cdot \rightarrow\end{array}\right\} \text{non-radical products}$$

The free radical $A^\cdot$ derived from the antioxidant is stabilised by resonance and so does not participate in the propagation steps.

Chain-breaking antioxidants used in the food industry are phenols, e.g.

Table 1.7 Lipid oxidation flavours

	Flavour thresholds (ppm) in:		
	Water	Milk	Paraffin oil
Alcohols			
Ethanol	200		
1-Propanol	45		
1-Butanol	7.5	0.5	
1-Hexanol	2.5	0.5	
1-Heptanol	0.52		
1-Nonanol	0.086		
1-Decanol	0.18		
1-Penten-3-ol	3.0	3.0	4.2
1-Octen-3-ol	0.001	0.01	0.0075
Aldehydes			
Hexanal	0.03	0.05	0.6
Heptanal	0.03	0.12	0.04
Octanal	0.047	0.46	0.6
Nonanal	0.045	0.22	0.32
Decanal	0.007	0.24	1.00

Selected from Forss.[9]

butylated hydroxyanisole (BHA), butylated hydroxytoluene (BHT), butylated hydroquinone (TBHQ) and propyl gallate. Such compounds lose their efficiency at high temperatures because the hydroperoxides formed as above break down.

There are also preventive antioxidants, which act by reducing the rate of chain initiation. Metal inactivators, which coordinate with metal ions capable of catalysing chain initiation, include citric, phosphoric and ascorbic acids. Some preventive antioxidants can absorb radiation without forming radicals. Carbon black, phenyl salicylate and hydroxybenzophenone are examples of UV deactivators.

Finally there is synergism, which is the effect obtained when two of these stabilisers are used together. The mixing of the two has a much better effect than either of the stabilisers alone. If a chain-breaking and a preventive antioxidant are mixed both initiation and propagation are suppressed.

In biological tissues, it is suggested that vitamin E and vitamin C act together to remove the free radicals generated in polyunsaturated fatty acid radicals. The trend towards 'natural' antioxidants has led to the use of tocopherols (vitamin E) along with ascorbic acid (vitamin C) and phospholipids which show intense synergism.

In this chapter an outline of some of the major chemical reactions has been given, but no attempt has been made to cover all of the more sophisticated elements of current research, e.g. superoxide dismutase, carbon NMR and mass spectrometric identification of hydroperoxides.

Some of these additional points can be gleaned from the publications suggested under Further reading.

References

1. Hamilton, S. and Hamilton, R. J. (1972). In *Materials and Technology Encyclopaedia*, Vol. 5, (ed. J. H. De Bussy), Longman, Amsterdam, pp. 219-25.
2. Hilditch, T. P. (1941). *Industrial Fats and Waxes*, 2nd edn, Bailliere, Tindall and Cox, London.
3. Hilditch, T. P. and Williams, P. N. (1964). *Chemical Constitution of Natural Fats*, 4th edn, Chapman & Hall, London.
4. Gunstone, F. D. (1967). *Introduction to the Chemistry and Biochemistry of Fatty Acids and their Glycerides*, 2nd edn, Chapman & Hall, London.
5. Swern, D. (ed.) (1979). *Baileys Industrial Oil and Fat Products*, Vol. 1, 4th edn, Wiley, New York.
6. Robards, K., Kerr, A. F. and Patsalides, E. (1988). *Analyst*, **113**, 213.
7. Frankel, E. N. (1980). *Prog. Lipid. Res.*, **19**, 1.
8. Kinderlerer J. L. (1986). In *Spoilage and Mycotoxins of Cereals and Other Stored Products*, (ed. B. Flannigan), C. A. B. International, Slough, UK.
9. Forss, D. A. (1972). Odour and flavour compounds from lipids. In *Progress in the Chemistry of Fats and Other Lipids*, Vol. 13 (ed. R. T. Holman), Pergamon Press, Oxford, pp. 177-258.

Further reading

Shultz, M. W., Day, E. A. and Sinnhuber, R. O. (eds) (1961). *Symposium on Foods: Lipids and their Oxidations*, AVI, Westport CT, USA.
Lundberg, W. O. (ed.) (1961-62). *Autoxidation and Antioxidants*, Vols. I and II, Interscience, New York.
Simic, M. G. and Karel, M. (eds) (1980). *Autoxidation in Food and Biological Systems*, Plenum Press, New York.
Vliegenthart, J. F. G. (1979). *Chem. Ind.*, 241.
Chan, H. W. S. (ed.) (1987). *Autoxidation and Unsaturated Lipids*, Academic Press, New York.
Min, D. B. and Smouse, T. H. (eds) (1985). *Flavor Chemistry of Fats and Oils*, American Oil Chemists' Society, Champaign, Illinois, USA.
Min, D. B. and Smouse, T. H. (eds) (1989). *Flavor Chemistry of Lipid Foods*, American Oil Chemists' Society, Champaign, Illinois, USA.
St. Angelo, A. J. (ed.) (1992). *Lipid Oxidation in Food*, American Chemical Society, Washington, DC, USA.

2 Measurement of rancidity
J. B. ROSSELL

2.1 Introduction

Rancidity in processed food is becoming increasingly important for two major reasons. There is commercial pressure for manufacturers to produce food at a smaller number of centralised locations to provide economies of scale. This in turn necessitates longer distribution runs, often via regional warehouses and thus requires longer shelf-lives. In addition, public awareness of nutritional issues is creating a demand for foods with higher levels of polyunsaturated oils, and reduced levels of food additives such as synthetic antioxidants. Foods with higher concentrations of polyunsaturated fatty acids will, of course, oxidise more quickly, with more rapid development of rancid off-flavours in consequence, especially if it is also necessary to remove antioxidants.

The measurement of rancidity in oils and fats can be carried out by both physical and chemical means. These should, however, be distinguished from methods that measure the resistance of an oil or fat to rancidity, rather than the rancidity itself.

Three types of rancidity need to be taken into consideration, namely hydrolytic rancidity, oxidative rancidity and the less well-known ketonic rancidity. *Hydrolytic rancidity* is caused by hydrolysis of the triglycerides in the presence of moisture, and gives rise to the liberation of free fatty acids (FFA). These free fatty acids are particularly troublesome in the lauric oils as the fatty acids liberated (namely capric, lauric and myristic) have strong off-flavours. *Oxidative rancidity* is caused by oxygen attack on the fat, with the development of oxidised products and associated off-flavours. The third type of rancidity, *ketonic rancidity*, occurs when fungal attack on foods, in the presence of limited amounts of oxygen and water, leads to liberation of short-chain saturated fatty acids. These then undergo a β-oxidation to produce a range of homologous even-carboned methyl ketones and aliphatic alcohols.

In all three cases, rancidity is a question of off-flavour. If an oil tastes rancid, then it is rancid. The first line of measurement is therefore by taste panel and subjective evaluation of an oil. The organisation of taste panels, in order to overcome the problems of subjective judgement and personal preference, is a subject in its own right. The organisation must be carried out carefully if the results are to be reliable, and the reader is advised to turn to specialist information on this topic.[1,2]

There are, of course, several reasons for wanting to measure rancidity. A manager in the food industry needs an effective method of quality control for the products leaving his factory. Food manufacturers want to ensure that oil or fat reaching their factories is in a good condition before it is combined with other ingredients in the manufacture of finished food products. A third area in which measurements are needed is in the systematic study of rancidity development. This may be where research or development work is striving to increase the resistance towards rancidity of a food, or where a manufacturer is comparing two candidate oils for his product and needs factual data to help decide his purchasing requirements. It is in this last area that the measurement of rancidity itself and the measurement of resistance towards rancidity become confused. A product may be placed on shelf-life test and periodically assessed to determine the time taken for rancidity to develop. The method of testing used is then one of true-rancidity measurement, but the results are used to indicate the resistance of an ingredient towards rancidity. Alternatively, an accelerated test may be used to predict the resistance of the product towards rancidity and in this case the rancidity itself is not measured. The topic of measuring the resistance of a product towards rancidity will be considered in due course.

Although taste assessment is in the long run the most reliable method, because it measures what the customer actually perceives, it is difficult to ascribe numerical data to it and thus make comparisons with results obtained on a different date or in a different location. It can also be expensive and cumbersome to carry out. It is for these reasons that many people look towards physical and chemical methods for rancidity measurement.

2.2 Measurement of hydrolytic rancidity

Hydrolytic rancidity is a problem mainly encountered in products based on lauric oils such as palm kernel or coconut. The rancidity is due to the liberation of free fatty acids from the parent oils, which therefore comprise large amounts of capric, lauric and myristic acids. These acids have a distinct soapy flavour, which is why hydrolytic rancidity is often referred to as soapy rancidity. The acids have lower flavour threshold values than the longer chain fatty acids found in other oils and fats, such as palm oil, soyabean oil or beef tallow, flavour thresholds of 0.07% and 0.02% having being reported[3] for lauric and capric acids, respectively. The flavour thresholds for longer chain acids are higher and are seldom encountered. In my experience, the problem is not often noticed in a food comprising a fat having these low free fatty acid levels, as other flavour ingredients mask the very slight favour of the free acids at the threshold value. A problem is encountered, however, when a lauric fat contains about 0.2% free acid. In a lauric substitute chocolate coating containing about 30% fat, problems begin to arise when the fat component in the coating contains about 0.5 or 0.6% free fatty acid.

Hydrolytic rancidity is generally caused by a combination of enzymes and moisture. It is worth remembering the molecular weight relationships of the reacting species during the development of hydrolytic rancidity. Equation (2.1) shows the liberation of lauric acid from fat containing as little as 0.1% moisture:

$$\underset{(\text{mol. wt, 600})}{\text{lauric fat}} + \underset{(\text{mol. wt, 18})}{0.1\% \text{ H}_2\text{O}} \xrightarrow[\substack{\text{e.g. lipase} \\ \text{enzyme} \\ \text{mould, yeast}}]{\text{catalyst}} \text{diglyceride} + \underset{(\text{mol. wt, 200})}{1.1\% \text{ lauric acid}}$$

(2.1)

This equation shows the amount (1.1%) of free fatty acid liberated when 0.1% moisture is fully reacted, and although complete reaction can take a long while, it is nevertheless evident that only small amounts of moisture are necessary for the development of sufficient lauric acid for an off-flavour to be perceived.

Where hydrolytic rancidity is suspected, there are therefore three factors that we can measure, namely moisture, catalytic agent and the free fatty acid level. The moisture and catalytic agent are, of course, the factors causing the hydrolytic rancidity, whilst the free fatty acid, if present, is the effect of the hydrolysis. Moisture may be measured by total volatile matter, and in general this author prefers the British Standard Method in BS 684 Section 1.10. An alternative is the Karl Fischer titration in BS 684 Section 2.1. The Karl Fischer titration[4] is, of course, most attractive with products that would give misleading answers during any heating to evaluate the total volatile matter, for example products containing appreciable quantities of volatile flavour components. It is also worth remembering that the free acids derived from lauric oils are themselves volatile, and the oven drying method in BS 684:1.10 is not therefore applicable to lauric oils; this is another reason for choosing the Karl Fischer method.

Hydrolysis seldom takes place in the absence of a catalyst, and this is usually a free lipase or esterase enzyme. Lipases and esterases are conveniently determined by the indoxyl acetate test, first introduced by Purr,[5,6] and subsequently developed at the Leatherhead Food RA.[7] In this test, a filter paper impregnated with indoxyl acetate is exposed to a paste of the test sample in a buffer solution. The presence of any lipase in the sample causes the liberation of indoxyl radicals, which oxidise and condense to form indigo blue on the paper during subsequent incubation at 37 °C. The blue coloration is easily seen, and can provide a semi-quantitative comparison under standard conditions.

There are sometimes interferences from extraneous chemicals in the indoxyl acetate test. For instance, some fumigants used on nuts give a blue-gree colour even when no microbiological activity is present. In such cases it can be expedient to incubate a paste of the foodstuff with coconut oil, and measure the FFA of the oil over a period of time.

Milk powder can contain microorganisms from poor quality milk, which can produce heat-stable enzymes resistant to spray drying even though the organisms may have died. These enzymes may be difficult to detect, but the indoxyl acetate test, or incubation with an emulsion of trioctanoin and chromatographic examination of the reaction product for free octanoic acid will usually reveal any such spoilage enzymes.

There are, of course, other types of catalytic agent that may promote hydrolysis, one of which is a soap residue remaining after refining. Crude palm kernel oil contains a low level of free fatty acid, which may be removed by an alkaline wash. Any residual alkali or soap is removed by hot water washing and the product is then dried, bleached and deodorised. However, there are cases where traces of soap or alkali may remain. Soap has an alkaline reaction and can catalyse hydrolysis of the fat in the presence of moisture. Cocoa powder intended for use as an ingredient in drinking chocolate is frequently alkalised. If a lauric coating is accidentally manufactured with alkalised cocoa powder, any residual alkalinity in the powder may also catalyse subsequent hydrolysis. If either of these possibilities is suspected, fat should be extracted from the coating and the free alkalinity in the fat determined by the Wolff titration described in BS 684 Section 2.5. Levels of soap up to 20 mg kg^{-1} (measured as sodium laurate) should not cause any trouble, although values in the range 5–10 mg kg^{-1} are, of course, better. It is when the alkalinity rises to levels of 50 mg kg^{-1} or more that problems of soapy rancidity are likely.

As mentioned above, moisture and catalytic agent are the causes of hydrolytic rancidity; the effect is the development of free fatty acid. The amount of free fatty acid that has developed in a product is conveniently determined by direct titration of the extracted fat against alkali according to the method in BS 684 Section 2.10. There are also alternative colorimetric methods,[8,9] for example with Rhodamine 6G, but this author prefers the direct titration method. This is accurate down to levels of 5 mg kg^{-1} and, if the level of acidity is as low as this, there is no problem of hydrolytic rancidity.

Since moisture and catalytic agent cause hydrolytic rancidity in lauric fats, there are three ways of avoiding the problem: reduce the amount or activity of the water, sterilise all sources of lipase, and take steps to remove any other catalyst. If none of these measures is successful, change to a non-lauric fat. An alternative, and perhaps underexploited way of reducing the influence of water is to treat the fat with a moisture scavenger. Lauric fats are therefore often dosed with about 0.1% of lecithin, which reacts with water, itself becoming hydrated but reducing the amount of water available for reaction with the fat.

Hydrolytic rancidity can develop in foods other than confectionery coatings, a topic discussed in several interesting paragraphs by Minifie.[10] An extensive review of the subject has also been published by Hogenbirk.[11]

2.3 Measurement of ketonic rancidity

Ketonic rancidity is an often overlooked problem with foods containing shorter chain saturated fatty acids. It can be a cause of off-flavour in desiccated coconut, but can arise in other foods, such as butter. It has been extensively studied by workers[12,13] at Sheffield Hallam University, who established that, in the presence of limited amounts of water and oxygen, moulds such as *Eurotium amstelodami* can interact with the lauric fat, first liberating short-chain free fatty acids, which are then subjected to a β-oxidation, yielding two homologous series of compounds, namely methyl ketones and aliphatic alcohols, both with odd carbon chains. There were also small amounts of even-carbon chain ketones and alcohols.

As the first step in the development of this form of rancidity is hydrolysis of the short-chain fatty acids, it is closely related to the hydrolytic rancidity discussed above. However, since the methyl ketones and aliphatic alcohols have different odours, the symptomatic musty, stale note distinguishes it from normal lipolytic rancidity.

Measurement of ketonic rancidity therefore parallels that of lipolytic rancidity, viz. analysis for moisture, lipase and/or microbiological activity and free fatty acid content. In addition, the volatiles can be distilled off and analysed for the homologous series' of methyl ketones and aliphatic alcohols, both with odd-numbered carbon chains.

2.4 Measurement of oxidative rancidity

In comparison with hydrolytic rancidity, oxidative rancidity is a more complex subject, and one to which a considerably greater amount of work has been devoted.[14-16] It is clearly caused by oxidation of the fat, and takes place through several intermediates including hydroperoxides and peroxides, eventually leading to aldehydes and ketones, as well as other breakdown products. These secondary oxidation products, especially the aldehydes, have the off-flavours associated with rancid oils. The mechanisms and nature of the oxidation reaction, as well as methods of inhibiting or preventing it, are discussed in other chapters. However, as the first product formed by oxidation of an oil is a peroxide, or hydroperoxide, the most common method of measurement is the peroxide value (PV), which is reported in units of milliequivalents of oxygen per kilogram of fat.

2.4.1 Peroxide value (PV)

There are numerous analytical procedures for the measurement of the PV. In all cases the results and accuracy of the test depend on the experimental conditions, as the method is highly empirical. The most common methods are those based on the iodometric titration originally reported by Lea[17] and Wheeler,[18] which measure the iodine liberated from potassium iodide by the peroxides present in the oil. It has been contended[4,19] that the two principal sources of error in these methods are the absorption of iodine at unsaturated bonds in the fatty acids and the liberation of additional iodine from potassium iodide by oxygen present in the solution.

Other sources of error can arise,[4] but many of these were taken into consideration in the British Standard Method BS 684:2.14 (1990), ISO 3960 and AOCS Cd 8-53. A new area of doubt has arisen in recent years, however, and this relates to the choice of solvent. Until a few years ago chloroform was the tried and tested solvent universally used. However, it is toxic and has also been implicated with the so called 'Greenhouse Effect', and its manufacture discouraged by governments worldwide. Use of alternative solvents is a potential source of deviation, as described later.

Several alternative methods have been recommended for PV determination and these include a colorimetric method based on the oxidation of ferrous to ferric ion and the determination of the latter as ferric thiocyanate,[20] a variation in the iodometric method reported by Swoboda and Lea[21] in which the liberated iodine is converted into a blue starch-iodine complex, and the Sully method[22] in which the mixture is boiled. The American Oil Chemists Society (AOCS) standardised a procedure for PV determination in its method Cd 8-53, and this is applicable to all normal fats and oils. As with all other methods, it is highly empirical and any variation in procedure can lead to inaccuracies in the result. This official method was modified[23] to increase its sensitivity. The method apparently failed to measure low PVs, as a result of problems in the determination of the titration end-point. In the modification,[23] the titration step is replaced with an electrochemical technique in which the liberated iodine is reduced at a platinum electrode maintained at a constant potential. It is claimed that PVs ranging from 0.06 to 20 units (mequiv O_2/kg fat) can be accurately determined by this means. All solutions must be deaerated during this determination, as the presence of oxygen leads to further peroxide formation during the test.

The peroxide value is a good guide to the quality of a fat, and in the view of this author freshly refined fats should have PVs of less than 1 unit. Fats that have been stored for some period of time after refining may be found to have PVs of up to 10 units before undue off-flavours are encountered.

In the analysis of foods, fat is often extracted from the food, e.g. with petroleum ether, the solvent evaporated, and the PV measured on the recovered oil. This procedure can be a major source of error. The extraction is hot and is seldom carried out with total exclusion of oxygen, including deaeration of both food and solvent, and there are often traces of oxygen remaining during solvent removal. Peroxides can therefore be formed during this procedure, sometimes in excess of those originally present. Conversely, during solvent removal, high temperatures may be used to drive off the last traces of solvent. This can cause decomposition of peroxides. These two sources of error seldom compensate.

A much better procedure, which has been validated at the Leatherhead Food RA,[24] is to extract the oil from the food by the well-known 'cold methanol then chloroform' method of Bligh and Dyer.[25] The total volume of chloroform solution resulting from this extraction method is noted, and aliquot portions evaporated to give the concentration of fat in the solution, and thus, if needed, the amount of fat extracted. Different aliquots of the chloroform solution are then used for direct determination of the PV, there being no need to remove the chloroform as this is used in the test itself, thus avoiding any heating during extraction or solvent removal.

All of the work discussed above relates to the use of chloroform to dissolve the fat and, as mentioned earlier, use of chlorinated hydrocarbon chemicals of this class is being discouraged. Alternative solvents have therefore been sought. The American Oil Chemists Society Recommended Practice Cd 8b-90 (Revised 1991) specifies the use of an iso-octane plus acetic acid solvent, collaborative study results having been published in 1990.[26] Eight laboratories participated in this collaborative test, which evaluated peroxide value determination in three oils, linseed, soyabean and sunflower seed. Each laboratory oxidised the oil samples at 60 °C over a 60-day period leading to PVs in the range 20–100 for linseed oil, 1–110 for soyabean and 1–300 for sunflower seed oils. There was a good one-to-one relationship between the results for the two solvents, viz. iso-octane/acetic acid versus chloroform/acetic acid, agreeing with earlier results carried out on a smaller scale and in which lard, peanut (groundnut), maize, sunflower seed, soyabean and canola oils had been used. However, this method with iso-octane/acetic acid has not been collaboratively tested in the normal sense of the phrase, i.e. a test in which participating laboratories analyse identical samples to confirm (a) that the test is rugged, (b) that all laboratories produce an equivalent answer, and (c) to establish the experimental error and reproducibility of the test. This is in any case difficult with PV as the oil samples used in the test are likely to change, by further oxidation or by decomposition of the peroxides already present, in the time interval between preparation at the coordinating laboratory and analysis in the participating laboratories. This can be a big problem if international postal services are used for sample distribution or if custom authorities find a need to open and examine the samples.

Peroxide value measurements are of less use in assessing the quality of highly unsaturated oils, such as unhydrogenated fish oils. This is probably because the peroxides initially formed are themselves highly unsaturated and thus unstable and react quickly to form secondary oxidation products. The amount of peroxide oxygen therefore remains low relative to that in liquid vegetable oils or animal fats, even after extensive oxidation. In these cases, tests such as that for the anisidine value, or the thiobarbituric acid (TBA) test for malondialdehyde, are often of more value.

2.4.2 Anisidine value (AV)

The peroxides in an oxidised oil are, of course, transitory intermediates, which decompose into various carbonyl and other compounds. This decomposition will accelerate as the temperature is raised, and the PV may therefore be reduced by heating the oil in the absence of air or oxygen. All crude oils are heated in the absence of oxygen during the bleaching and deodorisation stages of refining. A badly oxidised oil may therefore be processed to give a product with a low PV and an acceptable, but perhaps inferior, flavour. Since the oil was damaged it will have a lower level of natural antioxidants present. It may be impossible to produce the very bland flavour of a non-oxidised oil as it will contain non-volatile carbonyls that still have a flavour contribution.[27] It is also likely that the peroxide decomposition products may catalyse further oxidation, or, as proposed by Holm et al.[27] decompose or react further, giving rise to new off-flavour compounds.

Holm et al.[27] appreciated this problem in 1957, and developed a method for the qualitative assessment of aldehydes in a fat by reaction with benzidine acetate. Heating and strong acids were avoided to minimise further reactions of any peroxides present. This method did not attempt to give a numerically correct value for the content of carbonyl compounds, or to give a direct measure of the taste of fat, or its stability. Not all carbonyls have an off-flavour, and those that do occur vary considerably in type and intensity of taste.[28]

This method of assessment was found to be extremely useful, but was criticised on the grounds of the known carcinogenicity of benzidine. It was therefore replaced by the related anisidine value test,[29,30] an analytical procedure standardised by the International Union of Pure and Applied Chemistry (IUPAC) in method 2.504. The AV is defined as 100 times the absorbance of a solution resulting from the reaction of 1 g of fat or oil in 100 ml of a mixture of solvent and p-anisidine, measured at 350 nm in a 10-mm cell under the conditions of the test.

In the determination of AV the fat is treated with p-anisidine reagent in iso-octane solution and the level of reaction products assessed spectrophotometrically at 350 nm. The test estimates the level of aldehydes, principally 2-alkenals, present in the oil. As the absorbance maximum shifts toward a longer wavelength with increasing unsaturation, and as the colour intensity is greater with 2:4-dienals than with 2-enals, the absorption maximum varies from oil to oil, and so does the intensity of absorption of the complexes. The value obtained is therefore, strictly speaking, only comparable within each type of oil.[29]

It is important that the sample and reagents are dry. In particular, the glacial acetic acid used in the preparation of the reagent solution must be anhydrous; if there is any doubt about this, the quantity of water may be determined by the Karl Fischer method,[4] and any traces of water removed by careful treatment with acetic anhydride.

The AV is calculated to the nearest whole number using equation (2.2):

$$\text{AnV} = \frac{25(1.2\,A_2 - A_1)}{M} \qquad (2.2)$$

where A_1 is the absorbance of the solution of fat or oil, A_2 is the absorbance of the reaction product of the fat or oil with p-anisidine, and M is the mass in grams of the fat or oil present in the 25 ml of test solution.

In this version of the test, the factor 1.2 in the equation arises from the fact that the solution containing the fat or oil together with the p-anisidine is diluted by the addition of the anisidine reagent, whilst the solution of the test sample itself is not diluted. Readers should be careful not to confuse the above equation with that appearing in other forms of the test[31] in which the unreacted test sample is diluted with solvent, and the factor 1.2 does not therefore arise in the calculation of the result. The AV test is particularly useful for abused oils with low PVs such as frying oils, and in evaluation of the totox value, described below.

2.4.3 Totox value

The AnV is often used in conjunction with the PV to calculate the so-called total oxidation value, or totox value: totox value = 2 PV + AV. In this equation the PV is given in double weighting, as Holm and Ekbom found in experiments[29] that, when an oil was heated at 200 °C under vacuum, 1 PV unit decomposed to give an increase of 2 AV units. Patterson[32] has rationalised this by pointing out that peroxides have two oxygens per molecule whilst aldehydes have only one.

As a rule of thumb for good oils, the AV should be less than about 10. In a few cases there may be an appreciable PV together with an appreciable AV and, as this often indicates the onset of progressive deterioration, this author uses the more stringent rule of thumb that in these cases the totox value should be less than 10. The totox value is often considered useful in that it combines evidence about the past history of the oil, in the AV, with that of the present state of the oil, in the PV. It can also be of use in abused oils, such as frying shortenings, and in oils likely to oxidise quickly for other reasons, e.g. the presence of pro-oxidants. The totox value drops when an oil is refined, owing to simultaneous reductions in both PV and AV. In fact, the totox value should fall by about half during both bleaching and deodorisation. Measurement of the totox value can therefore be an important practical step in monitoring a refinery process, and can be used to identify faults, such as an air leak.[26,29]

2.4.4 Thiobarbituric acid (TBA) test

The TBA test is another empirical method frequently used for the detection of lipid oxidation. The test relates to the level of aldehydes present in the oil in a similar way to the AV test. In particular, it has been contended by Sinnhuber *et al*.[33] that thiobarbituric acid reacts specifically with malondialdehyde to give a red chromogen, which may then be determined spectrophotometrically. The proposed reaction is shown in Figure 2.1. However, it is possible that other aldehydes react with thiobarbituric acid to give red chromogens,[34] and that non-extractable lipids, urea, sugars, oxidised protein or other oxidised material present in foods can react to give coloured products.[35-37] Nevertheless, this can give an attraction to the test in that it can be carried out on whole foods, and may pick up oxidation damage to materials other than the extractable triglyceride fats themselves.

Where particular oxidation products are being studied, choice of analytical wavelength can distinguish chromogens from different reaction species. Thus unsaturated alka-2,4-dienals and alk-2-enals produce the red chromogen absorbing at 530 nm, while aldehydes in general produce

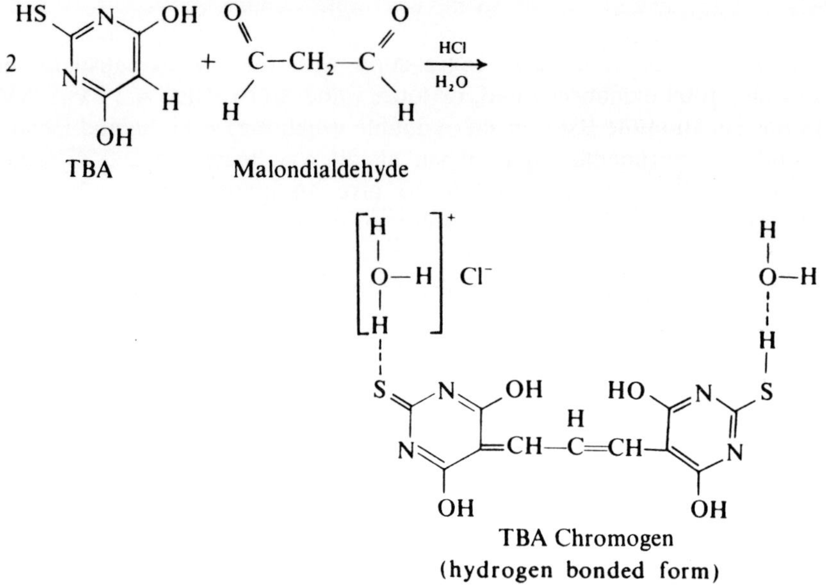

Figure 2.1 Proposed TBA reaction.

a yellow chromogen absorbing at 450 nm.[38] The separate chromogens have also been analysed by high-performance liquid chromatography (HPLC) prior to measurement at 546 nm.[39]

There are two basic methods of carrying out the test, either directly on the food product followed by extraction of the coloured pigment, or by steam distillation of the food and reaction of the distillate with the reagent. This author prefers the version of the test published in Pearson's Chemical Analysis of Foods.[40] In this test 10 g of fatty food are macerated with 50 ml of water for 2 min and the mixture is then washed into a distillation flask with 47.5 ml of distilled water; 2.5 ml of 4 N hydrochloric acid are added to the mixture, together with an anti-foaming agent and a few glass beads. The flask is heated until 50 ml of distillate are collected, which takes about 10 min from the time boiling first commences; 5 ml of this distillate are pipetted into a glass-stoppered tube, and 5 ml of 0.2883% (w/v) TBA solution in 90% glacial acetic acid are added. The tube is stoppered, shaken well and heated in a boiling water bath for 35 min. A blank tube is similarly prepared using 5 ml of water and 5 ml of the reagent. Both tubes are cooled in water for 10 min and the absorbance (D) is measured against the blank at 538 nm using 10 mm cells. The TBA number is then calculated in milligrams of malondialdehyde per kilogram of sample, which is equal to 7.8 times D. Several papers[19,41,42] have appeared reviewing and criticising the TBA test, which is said to be of little value with frying oils[43] but useful in measuring early stages of rancidity in liquid vegetable oils, lard and cooking fat,[44] and flesh foods.[40]

2.4.5 Kreis test (rancidity index)

The Kreis test[45] has the advantage that it is rapid, and also gives an indication of incipient rancidity. In fact it has sometimes been criticised[14] on this score. It was one of the first tests used to evaluate the oxidation of fats, and involves the production of a red colour when phloroglucinol reacts with oxidised fat in acid solutions. Early literature reports[46–48] that the compounds responsible for the Kreis colour reaction are epoxy aldehydes or their acetals.

The test may be carried out in a qualitative and also a quantitative form, even though fresh samples free from rancidity sometimes show a slight colour when reacted with the Kreis reagent. The method is standardised in BS 684 Section 2.32 (1979; reconfirmed 1985). Briefly, it involves reaction of the fat sample or food extract with phloroglucinol in diethyl ether solution. The products are then extracted with hydrochloric acid, and a red aqueous solution is obtained if the material is rancid. This red colour is then measured with a Lovibond colorimeter in a 1-inch glass cell. The results are reported in red units on the Lovibond scale. As noted above, a very faint red colour does not necessarily denote rancidity,[19] but a colour of up to 3 red units is considered to indicate incipient rancidity. If a colour reading between 3 and 9 red units is obtained, it is considered to indicate rancidity towards the end of the induction period. Colour readings of over 8 red units are considered to indicate definite rancidity. It should be noted that some food additives, especially vanillin, can interfere with this test, and it is therefore recommended that a separate test for vanillin should be carried out if its presence is known or suspected. Cavazzana has correlated spectrometric measurements at 546 nm with organoleptic evaluations.[49]

2.4.6 Other chemical methods

Several other chemical tests have been suggested for the measurement of rancidity in fats. These include the American Oil Chemists Society Method Cd 9–57 for oxirane oxygen in epoxidised products, and a procedure[50,51] based on the formation of carbonyl 2,4-dinitrophenyl-hydrazones in the presence of trichloracetic acid. This method has the advantage that it attempts to measure carbonyl compounds that contribute to the rancid flavour, unlike the PV method, which is limited to the early stages of oxidation. Peers and Swoboda[52] suggested the determination of octanoate, a residue of oxidation, which should remain bonded to the glycerol backbone of the triglyceride molecule even after the initial peroxides have decomposed. Octanoate is therefore liberated by transmethylation to form methyl oxtanoate, which is then determined against a standard by GLC. The method is not applicable, however, to lauric or dairy butterfats, or blends with these, as these fats have initially high and variable levels of endogenous octanoate.

2.4.7 Physical methods

Several spectrometric techniques have been suggested for measuring rancidity in fats.

When linoleic acid is oxidised to form linoleic acid hydroperoxide the double bonds in the fats became conjugated. This is also the case when linolenic acid becomes oxidised to form linolenic acid hydroperoxide, and is the basis of the AOCS lipoxidase method Cd 15-78 for the determination of essential fatty acids.[53-55] The conjugated acids so formed absorb ultraviolet light; linoleic acid hydroperoxide, and the conjugated dienes that may result from its decomposition, show an absorption band at about 232 nm, while the secondary oxidation products and particularly diketones show an absorption band at about 268 nm. Conjugated trienes show a triple absorption band, of which the principal peak is in the neighbourhood of 268 nm, together with a secondary peak at about 278 nm. The determination of the specific extinction in ultraviolet light is described in BS 684 Section 1.15 (1990). Factors for the conversion of the absorptions into levels of specific compounds are given in AOCS Method Cd 7-58 but, in view of the fact that absorption peaks of several compounds overlap in this UV range, these factors have to be used with caution.

Infrared spectroscopy has also been used for the measurement of rancidity, but IR is of most value in the recognition of unusual functional groups and in the study of fatty acids with *trans* double bonds. O'Connor[56] reported that the appearance of bands at about 2.93 μm is due to the formation of hydroperoxides, whereas the disappearance of a band at about 3.2 μm indicates the replacement of a hydrogen on a double bond with some other free radical. This may indicate polymerisation. It is also suggested that the appearance of additional bands at about 5.72 μm, which is equivalent to C=O stretching, indicates the formation of aldehydes, ketones or acids. Furthermore, an increase in the absorption band at 10.3 μm indicates the formation of additional *trans* double bonds, probably due to damage caused during the oxidation.

Refractive index measurements have been used to measure the stability of sesame/soyabean oil blends,[57] but the main attraction of refractive index measurements, or UV and IR spectrometry is in association with HPLC, as described below.

Several techniques based on chemiluminescence and bioluminescence have been reported,[16] the main problems with chemiluminescence having been the rather weak emission, and the need for sophisticated equipment. However, Miyazawa *et al*.[58] applied an improved chemiluminescence technique to the study of rancidity and shelf life of fish. They conclude that the method is versatile and rapid when coupled with oxidation of luminol in the presence of cytochrome c, and gave good sensitivity to fatty acid hydroperoxides. Methods based on fluorimetry[59] have also been applied, mainly as a means of measuring lipid oxidation in biological tissues and, in some cases,[58] analysis of purified fluorescent oxidation products was accomplished by IR evaluation of extracts separated by thin layer chromatography (TLC).

Several polarographic methods have been developed for the determination of oxidised lipids.[60,61] The fat is dissolved in a non-aqueous solvent and the oxidised components are reduced at a dropping mercury electrode.[60,61] A suitable reference electrode must be provided, via either a salt bridge or a sintered glass membrane. If benzene/methyl alcohol with dissolved lithium chloride is used as the non-aqueous medium, unsaturated acids may also be determined. The polarographic technique has the advantage that several species can be independently determined in the same reaction medium. New polarographic developments have enabled the dropping mercury electrode to be replaced by static electrodes and associated electronic equipment but, notwithstanding these advances, polarography seems to be a neglected area of research into methods of rancidity measurement.

Oxidation rather than rancidity has been followed in frying oils by physical measurements such as dielectric constant,[57,62] but care must be taken not to confuse increases in dielectric constant due to oxidation with those due to entrainment of food particles or moisture droplets, the latter having a very great effect.

2.4.8 *Chromatographic methods*

Liquid column chromatography has been used to separate polar and non-polar material in used frying oils,[62,63] but the main attraction is in HPLC since it offers the opportunity to measure concentrations of thermally labile peroxides,[64] hydroperoxides[65] and both volatile[66] and non-volatile[67,68] secondary oxidation products.

A major development in the gas chromatographic determination of rancidity is in the measurement of the volatile hydrocarbons released by the fat during the decomposition of the oxidised intermediates. Pentane is one of the main hydrocarbon gases released through thermal decomposition of the oxidised intermediates[68,69] and significant correlations of flavour scores with the evolution of volatile alkanes, in particular pentane, have been used to determine the rancidity of oils. AOCS Recommended Practice Cg 1–83 relates total volatiles to a flavour panel evaluation. Several approaches can be considered; in one of these the oil is injected directly on to a chromatographic column. The bulk of the oil remains at the start of the chromatographic column, but volatile components are carried through the column by the carrier gas.[70] One problem with this approach is that any decomposition of the oxidised intermediates will give volatile products and lead to an unstable chromatographic base line. Min[71] improved on the direct injection technique by injecting the oil into a U-tube for isolation of volatiles. The U-tube was connected to the GC to separate the isolated flavour compounds, enabling separation and determination of decadienals and other volatile components. Scholz and Ptak[66] used direct injection of cottonseed and other oils on to the column to determine pentane levels, as this was claimed to minimise sample decomposition. Correlations between flavour score and pentane release were established, but it was shown that a different relationship existed for each type of oil.

Several workers using direct injection have remarked[70,71] that column or pre-column life can be shortened by injection of the oil itself into the GC apparatus. However, in some work carried out in the author's laboratory,[72] a closely related technique was used to determine hexane solvent residues in oils and fats. The oil was injected on to a 10-cm glass pre-column packed with silanised glass wool. By changing the pre-column frequently, column life was preserved to the extent that over 200 determinations were carried out without any loss of sensitivity.

Several types of headspace analysis have been reported. In the most straightforward of these, the oil is heated in a closed tube; the volatiles collect in the headspace above the oil and are then analysed by GC. There are problems with this approach, however. For instance, it is appropriate to maintain a high temperature in order to drive the volatiles into the headspace, but there is also a requirement for a low temperature in order to minimise further reactions of the peroxides and secondary oxidation products in the oil. There is also uncertainty about the partitioning of volatiles between the vapour and the liquid oil.[16] In a potato chips–helium system at 140 °C the partition ratio for hexanal was 0.148,[73] a value that would be even lower for higher molecular weight aldehydes. In order to obtain suitable sensitivity, a volume of 2 ml headspace gas needs to be injected on to the column, which gives rise to a reduced septum life and risk of analyte loss through leaking septa. Various other problems with this form of headspace analysis have been reviewed.[72]

In an alternative headspace technique[74] the volatiles were steam distilled from aged potato chips, and pentanal and hexanal were then determined in the headspace gas by GC on a Carbowax 20 M column. The concentrations of both compounds provided a good measure of the quality of the chips.

Another attractive alternative to direct injection is to vacuum distil the volatiles into a cold trap. Blumenthal et al.[75] degraded frying oils by 'frying' moist cotton balls. Volatiles were distilled from the used oils at 90 °C and 0.05 Torr, and collected in traps cooled by solid CO_2 in acetone. The volatiles were collected and analysed by a GC fitted with a heated sniffer/collection port as well as a flame ionisation detector (FID). The GC peaks of the volatiles were numbered and, although not separately identified, the areas of several of the peaks correlated strongly with flavour evaluation of the used oils, each oil giving a different correlation.

An approach now becoming more popular with the commercial availability of porous polymer traps such as Tenax (R) is to purge the heated sample with an inert gas and trap the volatiles on a cool column of porous polymer. The column is then fitted to the GC, and the volatiles liberated by warming and analysed in the normal way. Singleton and Pattee[76] used this technique to measure volatiles produced from soyabean and groundnut oils oxidised in the presence of lipoxygenase enzyme. Pentane, pentanal and hexanal were the main volatiles produced in each case, and although different ratios of volatiles were produced from each oil, the ratios stayed constant even if the porous polymer column traps were stored for 70 h at ambient temperature prior to analysis. Dupuy et al.[77] used a related technique to measure concentrations of five volatiles in salad oils and shortenings.

These various headspace GC approaches can, of course, be applied to both oils and fatty foods, there being no necessity to separate or extract the fat from the food. However, in the case of more complex foods, difficulties in the identification and standardisation of the separate peaks in the presence of normal flavour volatiles can outweigh the advantages of not needing to separate the fat from the other food components.

2.5 Measurement of resistance to oxidative rancidity

There are several methods of measuring the resistance of a fat to oxidative rancidity and, as mentioned earlier, one of the easiest of these methods is to place typical fat samples on shelf-life tests. This is effective in comparing one type of fat against another, but is not of any practical use when comparisons between different batches of the same type of fat are to be made. In this latter case, it is quite clear that the information will be available only after the fat has gone rancid, which is too late for most practical applications. Several accelerated methods of testing have therefore been developed, three of which involve the use of elevated temperatures. It is appropriate to consider these three methods of testing, namely the Schaal oven test; the Sylvester test and its automated forms the FIRA–Astell and Oxidograph apparatuses; and the Swift test and its automated forms the Rancimat and OSI instruments.

2.5.1 Schaal oven test

The Schaal oven test for estimating resistance to oxidation involves heating 50–100 g of the sample in an open dish, and holding this in a thermostatically controlled oven until rancidity starts. The temperature at which the fat is held is either 63 ± 0.5 °C or 70 °C.[78] The fact that two different temperatures have been specified is an unfortunate confusion. The samples are then examined at regular intervals, for example daily or weekly, depending upon the keeping quality of the fat, and the condition of the fat is determined (this is done most easily by smell or taste). Alternatively, the course of the oxidation can be followed by determination of chemical factors such as the PV.

In the initial stages of the oxidation any antioxidants present in the oil will play their role and inhibit oxygen attack on the oil. This may be by removal of dissolved oxygen, in which case the antioxidant is an oxygen scavenger. Alternatively, the antioxidant may act by eliminating the free radicals generated.

Autoxidation in oils normally takes place through branching chain reactions, and the interruption of these, for example by the action of antioxidants, reduces the reaction speed considerably. However, in both of the above cases, the antioxidant is gradually consumed, after which oxidation of the oil becomes progressively more rapid. The initial stage of very slow oxidation is called the induction period (IP), and if the PVs measured during the Schaal oven test are plotted against time, the end of the induction period can be readily determined by a sudden upward turn in the graph (Figure 2.2). However, there is no guarantee that the value for the IP so determined will correspond with the life of the oil determined by smelling and tasting the samples.

The test has several weaknesses, which include inadequate control of the oxidation conditions and the subjective nature of smell and taste end-point measurements. If reasonably consistent results are needed, the taste tests should be conducted by qualified personnel working under standardised conditions.

A further problem with the Schaal oven test, and one that is common to all accelerated stability tests, is that of cleanliness of the apparatus. As fat autoxidation is autocatalytic, any residues of oxidised fat from a previous determination will catalyse oxidation of the test sample, and thus give a spuriously low result for its stability. Glassware should therefore be scrupulously clean, but cleaning agents containing transition metal ions, such as chromic acid, should not be used, as it is sometimes found that the glass surface can retain the transition metal ions,[79] and that

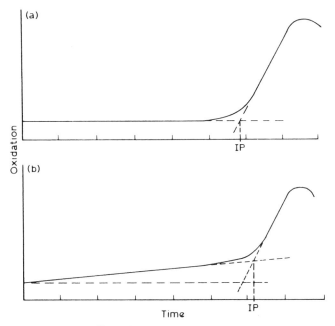

Figure 2.2 Examples of induction period (IP) curves.

these catalyse further oxidations. Glassware is most easily cleaned immediately after completion of the test, before the oil has a chance to polymerise and harden. Care should also be taken to ensure that the test sample is not contaminated by volatile material remaining in the oven from previous tests, or indeed stemming from parallel tests on samples with lower stabilities. Such volatile matter would comprise aldehydes and ketones, which would not only interfere with the organoleptic evaluation of the sample, but might also act as pro-oxidants. Needless to say, the inside of the oven should also be kept in a reasonably clean state, free from any oil spillages, as these will oxidise in the oven, giving rise to volatile decomposition products, jeopardising the reliability of any tests.

2.5.2 Sylvester test

In the original Sylvester test[80] a sample of the fat is placed in a closed vessel, which is heated to 100 °C in a thermostatically controlled bath and continuously shaken. The oxygen in the air above the oil reacts slowly with the oil as the latter becomes oxidised, leading to a reduction of pressure in the flask. This pressure drop can be monitored and a graph plotted showing the extent of oxygen uptake as a function of time. Oxidation is at first slow owing to the natural resistance of the oil to oxidation. When this is exhausted, however, oxidation becomes progressively more rapid. The graph of oxygen uptake against time therefore shows a distinct break at the end of the IP. Typical graphs are shown in Figure 2.2.

An improved Sylvester apparatus developed by Martin[81] and used for comparison of the stabilities of samples of fats extracted from cereals was equipped with an automatic recording facility, and was regarded[81,82] as preferable to the Swift test, the Schaal oven test and the ASTM oxygen bomb method. The experimental results were used for a comparison of the measured IPs with organoleptic assessments and PVs.

A problem with many of these systems was the need for continuous stirring throughout the duration of the test. Some fats are found to have induction periods ranging up to 200 h or even 400 h, and the provision of constant stirring for these long periods of time could not be adequately guaranteed. In addition, an enlarged and improved version of the apparatus, designed to enable the simultaneous assessment of four samples, led to an unnecessarily complicated and cumbersome recording system. Nevertheless, several workers, including Naggy et al.[82] noted that the oxygen absorption method correlated better with storage tests than did the results obtained by active oxygen methods such as the Swift test. This led the Leatherhead Food Research Association to put additional effort into the development of an automated version of the Sylvester test.[83] This was marketed commercially by the Astell Company and was commonly known as the FIRA-Astell apparatus.

2.5.3 FIRA-Astell apparatus

This apparatus was designed for the routine determination of oxygen absorption by fats and oils, by measurement of the pressure drop in a test flask. The apparatus was equipped with a pressure balance flask for each of the test flasks, and by coupling each of the flasks to a balancing diaphragm, which was in turn connected to an automatic recorder, the effect of changes in atmospheric pressure was eliminated. The apparatus was thermostatically controlled in the range 50–150 °C, and provided with constant stirring via magnetic followers.

The availability of the FIRA-Astell apparatus enabled more systematic testing of IPs. It was found that curves of different shapes were obtained, as illustrated in Figure 2.2, and some slight confusion arose over the best interpretation of such plots. In some cases the oxygen absorption is very nearly zero during the induction period and the graph therefore stays very close to the baseline. There is then a sudden departure as the rapidly increasing oxidation takes place, until most of the oxygen is absorbed. In many cases the peroxides and hydroperoxides thus formed slowly decompose, giving rise to secondary oxidation products, volatile at the temperature of the test (usually 100 °C), which lead to an increased pressure. This type of behaviour is illustrated in Figure 2.2(a). In other cases there is a steady change in pressure, leading, for example, to a slowly rising curve prior to the inflection. In cases such as this the author prefers to extrapolate the slowly rising part of the curve and take the point of intersection of this with a tangent drawn to the inflection of the curve. This is illustrated in Figure 2.2(b). This method gives a slightly longer induction period than would be obtained by extrapolating the tangent back to the chart baseline; it is considered to be more reliable, as it eliminates any chance changes in pressure due to causes other than oxidation.

It has been reported[80] that IPs for vegetable oils become better defined when oxygen is used in the headspace of the vessel. This is certainly the case when the oil already contains oxidised components, as these will lead to the evolution of volatile decomposition products and thus confuse the pressure drop due to oxygen absorption.

Although the FIRA–Astell apparatus was a considerable improvement on earlier techniques, it did have several drawbacks. In particular, the stirrer was still of insufficient reliability. This was mainly the case in busy laboratories routinely determining IPs on six samples simultaneously, some of which had long IPs overlapping those of other samples, so that it was difficult to turn off the instrument for maintenance or adjustment.

Leaks can also be a problem with this type of apparatus, as the basis of the test is pressure measurement. Leaks can occur from time to time in the rubber tube connections, especially if these become cracked or perished, and in the metal pressure transducers which may be attacked by the acidic oxidation products. Partly as a result of these problems, manufacture of the apparatus has now ceased.

Nevertheless, when the apparatus was well maintained, reliable results could be obtained, and a good deal of useful work was carried out at the Leatherhead Food RA with this apparatus.

IPs can range from less than an hour to several weeks, depending upon the resistance of the oil to oxidation at the temperature of measurement. In view of this, a series of tests was conducted in which the IPs of a series of oils were investigated at a series of different temperatures. Several typical results[84] are given in Table 2.1. Comparison of samples 2, 5 and 10, which are all samples of cocoa butter, shows that while sample 2 had the highest IP at 100 °C (indicating greater resistance to oxidation), sample 5 had the longest IP at 150 °C and sample 10 had the longest IP at 140 °C.

Similar comparisons can be made with the other samples. This strange relationship was not due to experimental error, as several samples had been evaluated at least six times at each temperature, showing quite low standard deviations, as illustrated in Table 2.2.

Table 2.1 Induction period of various fat samples at six temperatures (from Kochhar and Rossell[84])

Sample no.	Sample	Induction period[a] (h) at:					
		100 °C	110 °C	120 °C	130 °C	140 °C	150 °C
1	Hard soya 45	280, 300 mean = 290	178	70	26.8	13.2	5.6
2	Brazilian cocoa butter	278, 286 mean = 282	87.3	39.6	13.4	6.6	3.0
3	$N_{20}60$ + soya	255, 270 mean = 262.5	136	58	21.8	10.8	4.8
4	$N_{20}44$ + rape	213.5[b]	100.9[b]	42.7[b]	16.2[b]	7.9[b]	3.3[b]
5	Nigerian cocoa butter (A)	209, 213.7 mean = 211.4	87.3	39.5	12.6	7.4	3.9
6	Hard soya 36	199, 209 mean = 204	88	30	15.1	7.8	3.5
7	$N_{20}44$ + soya	193, 204 mean = 198.5	100.4	44	18	9.0	4.0
8	Non-deodorised cocoa butter	184, 176 mean = 180	75.3	32.5	12.6	7.0	3.4
9	Durkex 500	137, 139 mean = 138	59.2	31	7.2	5.2	2.6
10	Nigerian cocoa butter (B)	125.6, 124 mean = 124.8	89.5	35.8	14.8	7.8	3.8

Published with permission of the Leatherhead Food Research Association. [a]Time to inflection. Where no clear inflection could be discerned, the time for full-scale deflection was taken. [b]Mean of six values.

Table 2.2 Mean values, standard deviations and coefficients of variation of six IP determinations on each of four fats at six different temperatures (from Kochhar and Rossell[84])

Sample	Parameter[a]	Temperature (°C)					
		100	110	120	130	140	150
Wesco	$\bar{x}$	65.37	26.78	12.17	4.17	2.32	1.12
	s	±1.12	±0.69	±0.50	±0.05	±0.18	±0.08
	$100s/\bar{x}$	1.71	2.58	4.11	1.2	7.76	7.14
Beef oleo	$\bar{x}$	110.37	45.77	23.48	9.62	5.15	2.45
	s	±1.96	±1.81	±0.51	±0.44	±0.18	±0.1
	$100s/\bar{x}$	1.78	3.95	2.17	4.57	3.5	4.08
Groundnut fat[b]	$\bar{x}$	71.77	50.25	6.87	2.77	1.22	0.92
	s	±1.69	±1.07	±0.16	±0.15	±0.13	±0.10
	$100s/\bar{x}$	2.35	2.13	2.33	5.42	10.66	10.87
$N_{20}44$ + rape[c]	$\bar{x}$	213.45	100.88	42.65	16.20	7.93	3.32
	s	±2.52	±2.55	±0.67	±0.46	±0.25	±0.26
	$100s/\bar{x}$	1.18	2.53	1.37	2.84	3.15	7.83

Published with permission of the Leatherhead Food Research Association. [a]Mean = $\bar{x}$. [b]= hydrogenated groundnut oil. [c]= hydrogenated rapeseed oil with solid fat content at 30 °C measured by NMR of over 44%. Standard deviation = s. Coefficient of variation = $V = 100s/\bar{x}$.

Plots of the logarithm of the reciprocal of the IP against the reciprocal of the absolute temperature, i.e. 'Arrhenius' type plots, gave straight-line relationships with most oils. Four of these are illustrated in Figure 2.3, which goes a long way to explaining the strange results of comparison of IPs of cocoa butters in Table 2.1. It can be seen that the Arrhenius plots for the Brazilian and 'non-deodorised' cocoa butters cross, showing that the relevant reactions in the oxidation of these two fats have different activation energies.

2.5.4 The Oxidograph

As mentioned above, the FIRA–Astell apparatus had several drawbacks and it is no longer manufactured. Nevertheless, the approach is attractive, and has been shown[82] to relate better with storage tests than results obtained by other active oxygen methods. Furthermore, the oxygen uptake method gives valid results when the oil or fat contains volatile antioxidants such as butylated hydroxyanisole (BHA) or butylated

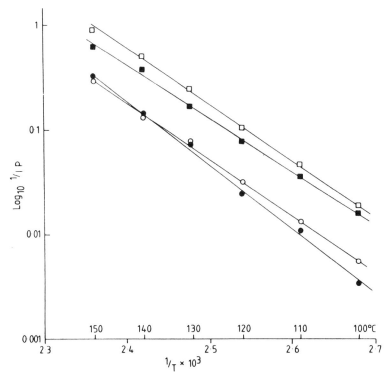

Figure 2.3 Arrhenius plots for IPs of four fat samples. ●, Brazilian cocoa butter; ○, non-deodorised cocoa butter; ■, 65% beef tallow/35% palm oil; □, palm/shea-based CBE.

hydroxytoluene (BHT). Partly for these reasons, the Grindsted Company in Denmark developed the Oxidograph apparatus, shown in Figure 2.4.

The basic apparatus consists of a thermostat; an electrically heated aluminium block in which six holes have been bored to accommodate six all-glass reaction flasks; a magnetic stirrer; six transducers, which are connected to the reaction flasks by narrow plastic tubing and which convert pressure changes to electrical signals; and an automatic recorder, which receives the electrical signals from the transducers and plots the pressure changes versus time.

In some tests at the Leatherhead Food RA, the Oxidograph was found to perform reliably and give induction period values in line with those anticipated from other work. In particular, it gave realistic values for oils or fats containing volatile antioxidants such as BHA and BHT, unlike other accelerated stability tests (described later) in which air is passed

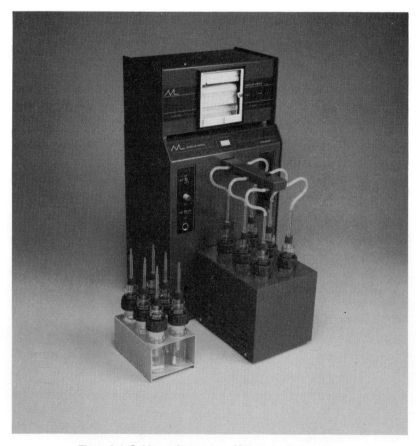

Figure 2.4 Oxidograph apparatus (Grinsted – Denmark).

through the oil, and in which the antioxidant is quickly swept from the oil at the elevated temperature of the experiment. It was also found that the Oxidograph gave reliable values with samples that were themselves volatile, such as an essential oil (orange oil) and certain emulsifiers. The Oxidograph also has simple, robust glass sample containers which are easy to clean, an important consideration in this type of work, as explained earlier. The glassware is also cheap to replace if vigorous cleaning leads to any breakages. Cleaning the glassware is easier in the Oxidograph than in tests that have a passing air stream as the limited supply of oxygen prevents excessive oxidation of the oil, which in turn prevents excessive polymerisation and hardening. Furthermore, retention of the volatile oxidation breakdown products helps soften any polymerised residues that do form. As the volatile components are retained, laboratory smells are minimal, an important aspect in laboratories where it is not always possible to open windows, and fume cupboards are not always available.

It was concluded that the Oxidograph is an excellent apparatus for rapid measurements of the IPs of oil and fat. It is advantageous for IPs of 24 h or less and can be used successfully for samples containing volatile antioxidants and for samples that are themselves volatile. Experimental advantages include simple design of glassware (which should therefore be cheap to buy, easier to clean and less breakable), no smell and only minor polymerisation of the substrates under test.

The apparatus is now manufactured by the Mikrolab Aarhus Company of Hojbjerg, Denmark. It has been described in a number of lectures at international symposia by Grindsted scientists.[85-87]

2.5.5 Swift test

The Swift test, or active oxygen method (AOM),[88,89] operates on a different principle from the Sylvester, FIRA–Astell and Oxidograph tests, as in this method air is bubbled through oil held at 98 °C. Samples of the oil are taken and the PV determined. The PV is then plotted against time and the IP determined by interpretation of the graph, as in the Sylvester test.

The method has been standardised in the Official and Tentative methods of the American Oil Chemists Society as Method Cd 12–57 and in BS 684 Section 2.25. These methods are entirely empirical, and it is therefore essential to adhere strictly to the described procedure in order to obtain reproducible results. In addition, it is advisable to state which of the texts was followed in the report of any results. The Swift test, or active oxygen method (AOM), proved very popular in some laboratories, and was developed into an automated version by Pardun and Kroll.[90]

2.5.6 Rancimat apparatus

This automated swift test works on a similar principle to the manual swift test, except that the effluent gases are led into a tube containing distilled water and the conductivity of the solution is measured between two platinum electrodes. The resulting signal is amplified and fed to a strip-chart recorder. Graphs are obtained showing the course of the oxidation.

Following additional development work by Zürcher and Hadorn,[91,92] a commercial apparatus, now called the Rancimat apparatus, has been marketed by Metrohm of Basle. It has several advantages over the FIRA–Astell and Oxidograph instruments, but also several disadvantages. These are reviewed in Table 2.3.

In the two approaches the IP graphs show similar features (Figure 2.2). In the Rancimat apparatus there is sometimes a slowly rising baseline to the graph, in this case caused by volatile components in the oil (e.g. free fatty acids of low molecular weight), which are carried over into the distilled water, causing an increase in the conductivity. A further similarity is that the graphs often reach a maximum and then fall again. In the FIRA–Astell/Oxidograph test this is due to the evolution of volatile components from the oil, causing an increase in pressure to compensate for the decrease resulting from oxygen absorption. In the Rancimat test, the constant stream of air eventually removes the dissolved acids from the distilled water, leading to a reduction in their concentration and an

Table 2.3 Features of FIRA–Astell and Rancimat tests

FIRA–Astell/Oxidograph	Rancimat/OSI
Limited oxygen supply	Always fresh supply of air/oxygen
Volatile components/products retained	Volatiles swept from system
No smell of oxidised oil	Volatile oxidation products can enter laboratory
Problems of mechanical reliability of stirrer and mechanical diaphragms	No stirrer needed
	External measuring system
Glassware easily cleaned and cheap to replace	Glassware difficult to clean and expensive to replace in Rancimat
Quiet in operation	Air pump can be noisy
Can be easily used for very long IPs provided there are no gas leaks	In long term use distilled water evaporates in hot air stream. The water trap should be maintained at low temperature in long IP runs to prevent loss of formic acid
Dust and condensing vapours are retained	Dust and condensing vapours are carried over in the air stream and block tubes and bubbler jets
Oxygen atmospheres give a sharper end-point but do not change IP	The passing air stream must contain no impurities that will change the conductivity of the water in the trap

increase in the electrical resistance of the water. The interpretation of IP curves can therefore suffer from this confusion in both cases. A discussion of curve interpretation has been published by van Oosten et al.,[93] working with a non-commercial (Unilever) form of this apparatus.

Several papers have appeared comparing the Rancimat apparatus with the 'manual' AOM method or Swift test, and reporting on its reliability,[94-96] and the apparatus has been shown to be reliable in many applications. De Man et al.[97] studied the composition of the gases evolved at the end of the IP. These comprised mainly formic acid with significant amounts of acetic acid and smaller quantities of propionic, butyric and caproic. The temperature of the water in the receiving jars was important, as water at a temperature of over 20 °C failed to retain the formic acid fully, which lead to errors in some cases.

In some work at the Leatherhead Food RA, FIRA-Astell (F) and Rancimat (R) IPs were compared. Some results are shown in Table 2.4. With most oils the Rancimat IP is slightly longer than the FIRA-Astell IP. A similar relationship was noted in some work comparing the Oxidograph with the Rancimat, and this has been interpreted as being due to the fact the FIRA-Astell and Oxidograph measure the uptake of oxygen, which must take place, in theory at least, just prior to the oxidation and to the end of the IP in the oil. In the Rancimat test, however, measurement is based on the evolution of volatile secondary products, which must take place after oxidation. In this respect it appears that the Rancimat slightly overestimates the resistance of an oil to rancidity.

In the presence of antioxidants such as BHA and BHT, the opposite relationship is found, however, as Rancimat IPs are much shorter than FIRA-Astell and Oxidograph IPs. This is clearly due to the fact that BHA and BHT are volatile at test temperatures of 100 °C or more, and

Table 2.4 Rancimat (R) and FIRA/Astell (F) induction periods (h) for some oils at 100 °C

Oil	Rancimat IP (R)	FIRA-Astell IP (F)	R/F
Palm kernel oil	12.8	8.9	1.43
Coconut oil	16.4	11.9	1.37
Palm oil	23.2	17.7	1.31
Hydrogenated rape (mpt 36/38 °C)	207	200	1.03
Soyabean oil	12.3	8.5	1.45
Tallow	9.5	6.8	1.40
Tallow + 100 mg/kg BHT	12.3	29.6	0.42
Tallow + 200 mg/kg BHA	23.5	86.7	0.27
Lard	10.6	8.2	1.29
Lard + 100 mg/kg BHA	15.7	19.7	0.80
Lard + 200 mg/kg BHT	15.0	32.9	0.46

are swept out of the oil by the passing air stream. The antioxidants are of course present at the beginning of the test, and so have some influence on the test result, as shown in Table 2.4. The full course of the oxidation will, however, depend more on the speed with which the antioxidants are swept out of the oil than on the resistance of the oil/antioxidant combination to oxidation.

Rancimat results are therefore of no value when volatile antioxidants such as BHA and BHT are present. However, several authors have published evaluations of BHA and BHT in which they have relied upon the results of Rancimat testing for their conclusions.[98–100]

The Rancimat 617 became widely used in the edible oil industry, and Rancimat induction periods are now written into a number of oil-purchasing specifications. The experience gained by industrial users over the years has been reviewed by the manufacturers, who introduced an upgraded model, the Rancimat 679. This is illustrated in Figure 2.5. The design of the second version alleviates some of the disadvantages listed in Table 2.3; for instance it now has two consoles connected by a 5-metre cable. This enables the reaction vessel console to be placed outside the working area of the laboratory or in a fume cupboard and smelly volatile reaction products that escape can thus be prevented from entering the laboratory.

Other improvements are the incorporation of an LCD display showing two lines each of 40 alphanumeric characters on which instrument para-

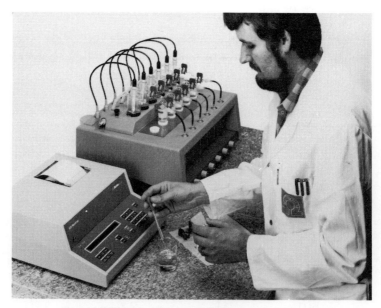

Figure 2.5 The recently upgraded Rancimat 679 apparatus.

meters, such as oil temperature, are presented. The instrument also has a solid thermostatically controlled heating block, and a waterproof polyester keyboard input for instrument settings. A microprocessor allows instrumental evaluation of the induction period, and printout of results as well as of the induction period graph for manual interpretation. At the time of writing, however, the author has not seen the 679 version in use, or received any reports from laboratories equipped with one. Recent publications on the Rancimat have reported the determination of oxidative stability of potato chips,[101] and temperature effects on the determination of oxidative stability.[102] In one paper,[102] attention is again drawn to the importance of cleaning the glass reaction tubes.

2.5.7 OSI apparatus

Following the success of the Rancimat apparatus in Europe, an alternative instrument was developed by the Archer Daniels Midland Company of Decatur, Illinois, USA. This was termed the OSI apparatus, as it used to determine the 'oil stability index', another name for the familiar induction period, except that it is now defined as the maximum in 'the second derivative of the conductivity with respect to time'.[103] This corresponds to the maximum slope of the IP curve (see Figure 2.2). As the OSI apparatus works on the same principle as the Rancimat, much of the discussion about the Rancimat applies here also. Several modifications have been introduced in order to overcome some of the problems workers found with the Rancimat. The OSI therefore uses disposable glass reaction tubes in order to reduce the need to clean away rancid oil thoroughly, and reduce the problem and cost of breakages. The electrical sensors are claimed to be rugged and long lived, with high mechanical stability. Up to twenty four samples can be analysed simultaneously at any two temperatures between 40 and 200 °C, and as the OSI interfaces with an IBM PC compatible computer, the curve differentiation used to obtain the maximum in the second differential can be obtained automatically. The physical layout of the OSI enables the easy introduction of a heated trap to collect any dust, mist or condensing vapour (e.g. high melting-point fatty acids), which might block bubble jets or contaminate the conductivity cell, the presence of dust causing significant problems in Rancimat evaluations of powdered fatty foods like coffee whiteners.

Several papers describing work with the OSI have been presented[104–107] and the apparatus has been approved by the American Oil Chemists Society in its Official Method Cd 12b-92.[103] It is illustrated in Figure 2.6, and manufactured by Omnion of Rockland, Massachusetts, USA. It has been recommended[102] that, in order to prevent a chaotic proliferation of interlaboratory communication and specifications over the temperature of testing, all results should be reported at, or factored to 110 °C. In the

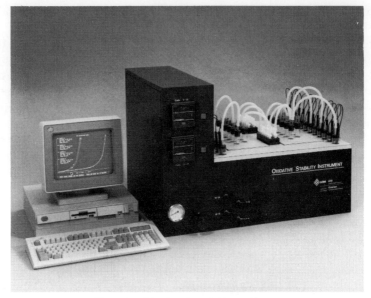

Figure 2.6 The Omnion 'OSI' apparatus.

opinion of the present author, this is a very worthwhile recommendation which could be followed in all methods of accelerated stability testing.

As with all oxidative stability tests, cleanliness of the apparatus is of the utmost importance. Several workers have referred to this, Pardun and Kroll[90] having prescribed a cleaning procedure that uses chromic acid. However, this is a procedure that this author does not recommend, since chromium ions can adhere to the glassware[79] and catalyse the oxidation of subsequent test samples. Standardised procedures have therefore been recommended. The American Oil Chemists Society official texts for the Swift test and the OSI[103] recommend a standardised cleaning procedure, which involves scrubbing with hot detergent solution, followed by rinsing and soaking in distilled water. Van Oosten *et al.*[93] report that a comparison of the results obtained by different laboratories on the same sample was of value only when the same method was used for cleaning the sample tubes, and the same graphical application of the curves applied.

Acknowledgements

I am indebted to the Director of the Leatherhead Food Research Association for permission to publish this paper, to Mrs A. Pernett for editorial help, and to Mrs R. A. Blaydon for arranging the typing.

References

1. Poste, L. M., Makie, D. A., Butler, G. and Larmond, E. (1991). revised edn., *Agriculture Canada*, Ottawa.
2. Steiner, E. H. (1966). *J. Food Technol.*, **1**, 41.
3. Forss, D. A. (1972). *Progress in the Chemistry of Fats and other Lipids*, (ed. R. T. Holman), Pergamon Press, Oxford, Vol. 13, p. 207.
4. Rossell, J. B. (1986). Classical analysis of oils and fats. In *Analysis of Oils and Fats*, (eds R. J. Hamilton and J. B. Rossell), Elsevier Applied Science Publishers, London and New York, p.1.
5. Purr, A. (1962). *Rev. Int. Chocolate*, **17**, 567.
6. Purr, A. (1965). *Nahrung*, **9**, 445.
7. Slack, P. T. (1987). *Analytical Methods Manual*, Leatherhead Food RA.
8. Chakrabarty, M. M., Bhattacharyya, D. and Kundu, M. M. (1969). *J. Am. Oil Chem. Soc.*, **46**, 473.
9. Anderson, M. M. and McCarty, R. E. (1972). *Anal. Biochem.*, **45**, 260.
10. Minifie, B. W. (1970). *Chocolate, Cocoa and Confectionery: Science and Technology*. J and A Churchill, London, pp. 412–14: 2nd edn. (1980), pp. 513–16.
11. Hogenbirk, G. (1987). *Manufacturing Confectioner*, May, 59.
12. Kinderlerer, J. L. (1986). *Int. Biodeterioration Suppl.*, **22**, 41.
13. Kellard, B., Busfield, D. M. and Kinderlerer, J. L. (1985). *J. Sci. Food Agric.*, **36**, 415.
14. Min, B. and Smouse, T. (1985). *Flavour Chemistry of Fats and Oils*, AOCS, Champaign, Illinois, USA.
15. Robards, K., Kerr, A. F. and Patsalides, E. (1988). *Analyst*, **113**, 213.
16. Duthie, G. G. (1991). *Chem. Ind.*, 42–4.
17. Lea, C. H. (1931). *Proc. R. Soc. London, Ser. B*, **108**, 175.
18. Wheeler, D. H. (1932). *Oil Soap*, **9**, 89.
19. Mehlenbacher, V. C. (1960). *The Analysis of Fats and Oils*, Garrard Press, Champaign, Illinois, USA.
20. Lips, A., Chapman, R. A. and McFarlane, W. D. (1943). *Oil Soap*, **20**, 240.
21. Swoboda, P. A. T. and Lea, C. H. (1958). *Chem Ind.*, 1090.
22. Sully, B. D. (1961). D. G. F. Einheitsmethoden C VIa, Wissenschaftliches Velagsgesellschaft, Stuttgart.
23. Fiedler, V. (1974). *J. Am. Oil Chem. Soc.*, **51**, 101.
24. Gillatt, P. and Rossell, J. B. (1987). *Leatherhead Food RA Res. Rep.* No. 579 (Members only).
25. Bligh, E. G. and Dyer, W. J. (1959) *Can. J. Biochem. Physiol.*, **37**, 911.
26. Berner, D. (1990). *Inform*, **1**, 884–6.
27. Holm, U., Ekbom, K. and Wode, G. (1957). *J. Am. Oil Chem. Soc.*, **34**, 606.
28. Frankel, E. N. (1983). *Prog. Lipid Res.*, **22**, 1.
29. Holm, U. and Ekbom, K. (1972). *Proceedings, International Society for Fat Research Congress*, Gothenburg, Sweden.
30. List, G. R., Evans, C. D., Kwolek, W. F., Warner, K. and Boundy, B. K. (1974). *J. Am. Oil Chem. Soc.*, **51**, 17.
31. Malaysian Oil Palm Growers Council (MOPGC) (May 1982). *Test Methods for Crude Palm Oil*, MOPGC, Kuala Lumpur, Method VI.
32. Patterson, H. W. B. (1989). *Handling and Storage of Oils, Fats and Meal*, Elsevier Applied Science, London and New York, p. 308.
33. Sinnhuber, R. O., Yu, T. C. and Chang, Y. T. (1958). *Food Res.*, **23**, 626.
34. Dahle, L. K., Hill, E. G. and Holman, R. T. (1962). *Arch Biochem. Biophys.*, **98**, 253.
35. Kohne, H. I. and Liversedge, M. (1944). *J. Pharmacol. Exp. Ther.*, **82**, 292.
36. Bernheim, F., Wilbur, K. M. and Fitzgerald, D. B. (1947). *J. Gen. Physiol.*, **31**, 195.
37. Butkus, H. and Rose, R. J. (1972). *J. Am. Oil Soc. Chem.*, **49**, 440.
38. Marcuse, R. and Johansson, L. (1973). *J. Am. Oil Soc. Chem.*, **50**, 387.
39. Bird, R. P., Hung, S. S., Hadley, M. and Draper, H. H. (1983). *Anal. Biochem.*, **128**, 240.

40. Kirk, R. S. and Sawyer, R. (eds) (1991). *Pearsons Composition and Analysis of Foods*, Longman Scientific and Technical, Harlow, Essex, UK, p. 642.
41. Kakuda, Y., Stanley, D. W. and van de Voort, F. R. (1981). *J. Am. Oil Chem. Soc.* **58**, 773.
42. Sans, R. G., Romero, V. Guillen, M. M. and Chozas, M. G. (1988). *Alimentaria*, **193**, 57.
43. Kim, D. H. and Maeng, Y. S., (1984). *Nonglim Nonjip*, **24**, 101.
44. Pokorny, J., Valentova, H. and Davidek, J. (1985). *Nährung*, **29**, 31.
45. Kreis, H. (1902). *Chem.-Ztg.*, **26**, 897.
46. Powick, W. C. (1923). *J. Agric. Res.*, **26**, 323.
47. Paton, S., Keeney, M. and Kurtz, G. W. (1951). *J. Am. Oil Chem. Soc.*, **28**, 391.
48. Lea, C. H. (1939). *Rancidity in Edible Fats*, Chemical Publishing, New York.
49. Cavazzana, O. (1980). *Rass. Chim.*, **32**, 135.
50. Henick, A. S., Benca, M. F. and Mitchell, J. H. (1958). *J. Am. Oil Chem. Soc.*, **31**, 88.
51. Gaddis, A. M., Ellis, R., Currie, G. T. and Thornton, F. E. (1966). *J. Am. Oil Chem. Soc.*, **43**, 242.
52. Peers, K. E. and Swoboda, P. A. T. (1979). *J. Sci. Food Agric.*, **30**, 876.
53. MacGee, J. (1959). *Anal. Chem.*, **31**, 298.
54. Kroll, J. and Mieth, G. (1977). *Nahrung*, **21**, 539.
55. Prosser, A. P., Shappard, A. J. and Hubbard, W. D. (1977). *J. Am. Oil Chem. Soc.*, **60**, 895.
56. O'Connor, R. J. (1956). *J. Am. Oil Chem. Soc.*, **33**, 1.
57. Yen, G. C., (1991). *Food Chem.*, **41**, 355.
58. Miyazawa, T., Kikuchi, M., Fujimoto, K., Endo, Y., Cho, S. Y., Usuki, R. and Kanada, T. (1991). *J. Am. Oil Chem. Soc.*, **68**, 39.
59. Dillard, C. J. and Tappel, A. L. (1973). *J. Am. Oil Chem. Soc.*, **8**, 183.
60. Lewis, W. R., Quackenbush, F. W. and de Vries, T. (1949). *Anal. Chem.*, **21**, 762.
61. Riccinti, C., Coleman, J. E. and Willis, C. O. (1955). *Anal. Chem.*, **27**, 405.
62. Yoon, S. H., Kim, S. K., Shin, M. G. and Kim, K. H. (1985). *J. Am. Oil Chem. Soc.*, **62**, 1487.
63. Waltking, A. E. and Wessels, H. (1981). *J. Assoc. Off. Anal. Chem.*, **64**, 1329.
64. Funk, M. O., Keller, M. B. and Levison, B. (1980). *Anal. Chem.*, **52**, 771.
65. Chan, H. W. S. and Levett, G. (1977). *Lipids*, **12**, 837.
66. Scholz, R. G. and Ptak, L. R. (1966). *J. Am. Oil Chem. Soc.*, **43**, 596.
67. Peers, K. E., Coxon, D. T. and Chan, H. W. S. (1981). *J. Sci. Food Agric.*, **32**, 898.
68. Ulberth, F. and Roubicek, D. (1992). *Fette Wiss-Technol.*, **94**, 19.
69. Pongracz, G. (1986). *Fette Seifen Anstrichm.*, **88**, 383.
70. Waltking, A. E. and Zmachinski, H. (1977). *J. Am. Oil Chem. Soc.*, **54**, 454.
71. Min, D. B. (1981). *J. Food Sci.*, **46**, 1453.
72. Downes, M. J. and Rossell, J. B. (1984). *J. Am. Oil Chem. Soc.*, **61**, 896.
73. Cabral, A. C., Orr, A., Stier, E. F. and Gilbert, S. G. (1979). *Package Dev. Syst*, **2**, 18.
74. Jeon, I. J. and Bassette, R. (1984), *J. Food Quality*, **7**, 97.
75. Blumenthal, M. M., Trout, J. R. and Chang, S. S. (1976). *J. Am. Oil Chem. Soc.*, **53**, 496.
76. Singleton, J. A. and Pattee, H. E. (1980). *J. Am. Oil Chem. Soc.*, **57**, 405.
77. Dupuy, H. P., Fore, S. P. and Goldblatt, L. A. (1973). *J. Am. Oil Chem. Soc.*, **50**, 340.
78. Williams, K. A. (1966). *Oils, Fats and Fatty Foods*, 4th edn, J and A Churchill, London.
79. Lang, E. P. (1934). *Ind. Eng. Chem. Anal. Edn.*, **b**, 111.
80. Sylvester, N. D., Lampitt, L. H. and Ainsworth, A. N. (1942). *J. Soc. Chem. Ind.*, **61**, 165.
81. Martin, H. F., (1961). *Chem. Ind.*, 364.
82. Naggy, J. J., Vibrans, F. C. and Kraybill, H. R. (1944). *Oil Soap*, **21**, 349.
83. Meara, M. L. and Weir, G. S. D. (1976). *Riv. Ital. Sostanze Grasse*, **53**, 178.
84. Kochhar, S. P. and Rossell, J. B. (1983). *Leatherhead Food RA Res. Rep.* No. 412 (Members only).

85. Larson, E. K. (1989). Methods for measuring antioxidation resistance. *Proceedings from Lipid Forum, 15th Scandinavian Symposium on Lipids*, Rebild Bakker, Denmark, pp. 562–65.
86. Winter, H. (1991). Methods for measuring the efficiency of antioxidants. Lecture at international conference *Antioxidants in Food*. Rebild-Bakker, Denmark.
87. Lausidsen, J. B. (1993). The Oxidograph for accelerated testing of antioxidants and the oxidative stability of oils and fats. *AOCS Annual Spring Meeting*. Anaheim, California.
88. Wheeler, D. H. (1932). *Oil Soap*, **9**, 89.
89. King, A. E., Raschen, H. L. and Irvin, W. H. (1933). *Oil Soap*, **10**, 105.
90. Pardun, H. and Kroll, E. (1972). *Fette Seifen Anstrich.*, **74**, 366.
91. Zürcher, K. and Hadorn, H. (1979). *Gordian*, **79**, 182.
92. Hadorn, H. and Zürcher, K. (1974). *Dtsch. Lebensm. Rundsch.*, **70**, 57.
93. Van Oosten, C. W., Poot, C. and Hensen, A. C. (1981). *Fette Seifen Anstrich.*, **83**, 133.
94. Läubli, N. W., Bruttel, P. A. and Schach, E. (1988). *Fette Wiss-Technol.*, **90**, 56.
95. Woestenburg, W. J. and Zaalburg, J. (1986). *Fette Seifen Anstrich.*, **88**, 53.
96. Läubli, N. W. and Bruttel, P. A. (1986). *J. Am. Oil Chem. Soc.*, **63**, 792.
97. de Man, J. M., Tie, F. and de Man, L. (1987). *J. Am. Oil Chem. Soc.*, **64**, 993.
98. Frank, J., Geil, J. V. and Freaso, R. (1982). *Food Technol.*, **36**, 71.
99. Pongracz, G. (1984). *Fette Seifen Anstrich.*, **86**, 455.
100. de Man, J. M. and de Man, L. (1984). *J. Am. Oil Chem. Soc.*, **61**, 534.
101. Barrera-Arellano, D. and Esteves, W. (1992). *J. Am. Oil Chem. Soc.*, **69**, 335.
102. Hassenhuetti, G. L. and Wan, P. J. (1992). *J. Am. Oil Chem. Soc.*, **69**, 525.
103. American Oil Chemists Society (1992). *Official Method* Cd12b-92.
104. Hill, S. E. and Perkins, E. G. (1992). Lecture to Midwest AOCS "82-92" Conference, Champaign, Il, USA.
105. Jebe, T., Matlock, M and Sleeter, R. T. (1993). *J. Am. Oil Chem. Soc.*, (in press).
106. Archer-Daniels Midland, *European Patent* 501682-A2.
107. Ayres, L. (1993). *Lipid Technol.*, **5**, 56.

3 Evaluation of oxidative rancidity techniques
B. J. F. HUDSON and M. H. GORDON

3.1 Introduction

Food scientists and technologists are frequently faced with having to make judgements about the quality of oils or fats which they handle, or products in which oil or fat contents are significant, both as received or stored and in relation to their end-uses. These judgements can be of two distinct, but closely related factors:

1. What is the quality of the product as it is received, in absolute terms? Is it acceptable, or has it already become 'rancid'?
2. If the product as received is of more or less satisfactory quality, how long, under defined storage conditions, is it likely to remain so?

The answer to the first question can be obtained directly from a simple sensory test. It is accepted for most oils and fats that a completely bland, tasteless product is ideal. There are exceptions—butter fat, olive oil, cocoa butter—but, broadly speaking, the problem for the taster is the detection of very small amounts of objectionable oxidation end-products. These are mainly aldehydes, but alcohols, ketones, carboxylic acids and lactones also play a part in off-flavours.

Alternatively, specific classes of end-products can be monitored by chemical analysis. These may be either those responsible or partly responsible for oxidation flavours, or other substances, possibly themselves of no great flavour significance but indicative of the last stage of the oxidation process.

The answer to the second question requires the development of predictive tests to forecast stability or shelf life. These can be accelerated deterioration studies or they can be measures of the levels of oxidation intermediates, such as peroxides. The formation of such intermediates, themselves without flavour, is a warning of rancidity to come, since they will inevitably break down in the course of time to produce off-flavour end-products.

Thus, while the first problem involves the determination of a chosen parameter (x), the second involves a time (t) parameter as well, so it may be expressed as dx/dt.

3.2 Sensory evaluation of rancidity

Rancid flavours are chemically very complex, since they are derived from any or all of the unsaturated fatty acids originally present in the oil and, as described in chapter 1, each of these can oxidise through several different mechanisms. Descriptive terms such as rancid, painty, beany, green, metallic, stale and so on are attempts to describe sensations arising from the presence of very small quantities of oxidation end-products in an oil or fat. Some specific, well-recognised aldehydes have distinctive characters. A few examples are given in Table 3.1.

It will be evident that these aldehydes are ultimately derived from linoleic or linolenic acids (see chapter 1). Most of them are disliked, through a few, such as *cis*-4-heptenal, can, at suitable concentrations, contribute positively to flavour.

In evaluating an oil or fat for suspected oxidative deterioration, trained tasters should participate in a standard triangular test and inconsistent results should be rejected. Following correct discrimination, flavour intensity should be based on a hedonic scale such as:

- Bland   0
- Suspicion of off-flavour   1
- Noticeable but very slight off-flavour   2
- More noticeable off-flavour   3
- Distinct off-flavour,   4
- Disagreeable off-flavour, rancid   5
- Markedly disgreeable off-flavour, very rancid   6

Tasters are provided with examples of oils scored 0 and 6 for comparison. Though there is considerable variation between individual tasters, reasonably consistent judgements are made with a panel of 30 experienced tasters. A mean flavour score (FS) is assigned on this basis to each oil sample.

In one example,[2] refined groundnut oil was heated at 100 °C in the

Table 3.1 Characters and threshold levels of selected aliphatic aldehydes[1]

Aldehyde	Threshold (ppm)	Character
cis-3-Hexenal	0.09	'Green bean'
trans-2-Hexenal	0.60	'Green'
cis-4-Heptenal	0.0005	Cream to tallow
trans-6-Nonenal	0.00035	'Hydrogenation'
trans-2-*cis*-6-Nonadienal	0.0015	Beany
trans-2-*trans*-6-Nonadienal	0.02	Cucumber
trans-2-*trans*-4-Decadienal	0.10	Stale frying oil

Table 3.2 Change in flavour score (FS) with duration of heating (Groundnut oil at 100 °C — 30 assessments)

Duration of heating (h)	Total FS	Mean FS	SD of FS
0	8	0.26	0.59
5	47	1.56	1.25
10	61	2.03	1.50
15	72	2.40	2.50
20	106	3.53	1.20

open for periods up to 20 h and the oil evaluated by 30 tasters at 5 h intervals. The results are shown in Table 3.2.

The AOCS has described a method for the flavour evaluation of vegetable oils by a flavour panel which uses a 10 point flavour grading scale with 10 being excellent and 1 repulsive.[3]

3.3 Estimation of oxidation artefacts

3.3.1 Estimation of specific off-flavours

It is evident from Table 3.1 that one of the most significant off-flavours, which is inevitably present in all oils containing linoleic acid, is *trans*-2-*trans*-4-decadienal. This fact is exploited in the determination of the anisidine value (AV), which relies on the condensation of *p*-anisidine with this and similar conjugated dienals to form coloured products absorbing in the UV with a maximum at 350 nm.

Of course there are many oxidation end-products, both aldehydic and otherwise, that will not respond to this determination, and some of them will undoubtedly have flavour significance. The more highly unsaturated vegetable oils, and marine oils, yield trienals to hexenals, which invalidate the test. As long, however, as dienals form a reasonably constant proportion of end-product off-flavours, AV, which is a straightforward determination, will be useful as an indicator of the actual state of rancidity of an oil.

3.3.2 Estimation of other end-products

Not all end-products have significant flavours, but some are nevertheless good indicators of rancidity. One of the oxidative breakdown patterns leads to the formation of volatile saturated hydrocarbons. Thus, the oxidation of oleic acid leads, *inter alia*, to octane, that of linoleic acid to pentane and that of α-linolenic acid to ethane. These can be released from the oil by heat and determined by gas chromatography. Since virtually all oils contain linoleic acid, which is subject to autoxidation, 'pentane values' have been employed to some extent in the evaluation of rancidity.[4]

'Headspace analysis' techniques, which utilise these features, are now applied quite widely and are particularly valuable in the routine evaluation of products such as powders and granules that are not conveniently handled by the more classical method.[5]

Headspace analysis of oils may be divided into two categories:

1. Headspace analysis with minimum decomposition of hydroperoxides.
2. Headspace analysis with thermal decomposition of hydroperoxides.

Analysis of volatiles in oil samples at 80–90 °C allows the detection of volatiles such as hexanal and pentane from $n-6$ fatty acids or propanal from $n-3$ fatty acids.[6] At temperatures above 90 °C, decomposition of hydroperoxides occurs sufficiently rapidly that the volatiles detected are dependent both on the hydroperoxides and the volatiles present.[7] Headspace analysis of oxidised oils is commonly performed by heating the oil at 180 °C, and this causes hydroperoxide decomposition in the sample vial, with the consequence that the volatiles analysed reflect both the volatiles present in the sample before heating and the volatiles formed by hydroperoxide decomposition. The total peak area of the volatiles produced under these conditions increases with storage of an oil, as shown in Figure 3.1. Although the volatiles produced from stored oils increase with peroxide value, the correlation is dependent on the oil under study.[8]

3.4 Measurement of induction periods

Classical autoxidation theory postulates three phases in the process. In the first, initiation, free radicals are gradually being formed, the rate of formation depending on the amount and degree of unsaturation of the component fatty acids. During this phase changes in the gross composition of the oil are hardly detectable, though it is claimed that very slight flavour deterioration, 'reversion', begins at this stage.

In the second phase, propagation, a critical level of free radicals has been reached and a relatively rapid chain reaction sets in. This phase is characterised by rapid absorption of oxygen, with the generation of peroxides. The third phase, termination, comprises the recombination of various species of free radicals, and thus has an effect in slowing down oxidation. Since, however, by this time substantial oxidation has already taken place, the termination stage has no effect on the development of rancidity.

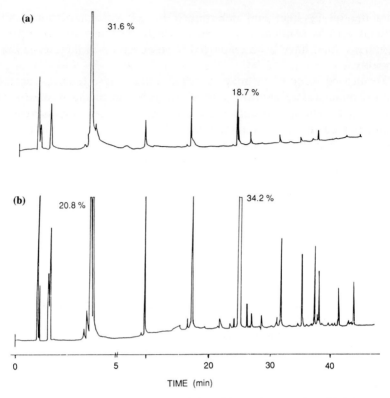

Figure 3.1 Chromatograms of headspace volatiles. (a) Total area 1.42×10^6, (b) total area 3.32×10^7.

Induction periods correspond closely with the onset of rapid absorption of oxygen and marked flavour deterioration. They can therefore be measured most directly by determining oxygen absorption as a function of time. In general, however, at ambient temperatures, induction periods are measured in weeks or months. It is therefore necessary to use accelerated tests, usually at 100 °C, or up to 140 °C for very stable oils and fats. Details of the methods in use are given in chapter 2.

Direct measurements of oxygen absorption can be made, but indirect measurement of, for example, peroxide values at predetermined time intervals, are more convenient. In the Schaal oven test (1938) samples are drawn daily for examination from a larger bulk sample maintained at 65 °C. In the Swift stability test (1933) a bulk sample is maintained at 98 °C and peroxide values (PV) determined at regular intervals, the induction period being taken as the time required for the PV to reach 20 milli-equivalents of oxygen per kg of oil.[9] Mehlenbacker (1942) developed a similar test, working at 110 °C.

Automated methods for induction period (IP) determinations are now available. The FIRA–Astell apparatus measures oxygen absorption by recording the rate of diminution of oxygen pressure in a closed system in contact with oil at 100 °C. IPs vary from about 1–2 h for very unstable fats such as lard, to 9 h for moderately stable oils such as refined soyabean oil, to 40 h for more stable oils such as palm oil and 100 h for very stable fats like cocoa butter. IPs can be markedly increased by hydrogenation or by the incorporation of various kinds of antioxidants.

An automated development, incorporated in the Metrohm Rancimat, monitors the progress of oxidation by measuring the change in conductivity of aqueous solutions of volatile oxidation products. The sharp rise in conductivity when the IP has been reached is due principally to the formation of volatile carboxylic acids as oxidation end-products.

IPs determined by these two automated methods generally correlate fairly well with each other, as can be seen from a few examples, quoted in Table 3.3. Since the generation of volatile carboxylic acids represents a later stage in the oxidation process than the absorption of oxygen, it is not surprising that IPs determined by Rancimat are somewhat longer than those determined by FIRA–Astell (see also section 2.5.6). It is also possible, therefore, to conclude that the rate determining step in the development of oxidative rancidity is the absorption of oxygen, not the decomposition of peroxides or subsequent transformations.

Induction periods are dramatically altered by the use of pro- and anti-oxidants or synergists. Incorporation of ionic copper at the parts per million level shortens them greatly, whilst, depending on any natural antioxidants which may already be present, additional primary antioxidants can greatly increase IPs. Synergists on their own have little effect, but in cooperation with primary antioxidants, whether naturally present or added, IPs are further enhanced.[10]

Oils which have been allowed to deteriorate by incubation at 100 °C exhibit both enhanced peroxide values and corresponding falls in IP. The quantitative aspect of this relationship is illustrated in Table 3.4.

Table 3.3 Induction periods (h) of oils determined by FIRA–Astell and Rancimat methods at 100 °C

	FIRA–Astell	Rancimat
Lard	1.4	2.1
Lard with 0.3% DPE[a]	1.7	2.5
Lard with 0.02% α-tocopherol	21.7	26.5
Refined soyabean oil (no citric acid addition)	6.7	7.9

[a]Dipalmitoyl phosphatidylethanolamine.

Table 3.4 Relationship between PV and IP in refined soyabean oil (citric acid added at the deodorisation stage) heated at 100 °C

Period of heating at 100 °C (h)	PV	IP (h)
0	0.6	13.7
1	1.2	13.2
2	1.8	12.2
4	2.2	10.8
6	3.2	10.3
8	3.8	8.5

3.5 Changes in iodine value or in component fatty acids

Iodine value (IV) is a measure of overall unsaturation and is still widely used to characterise oils and fats. Oxidative rancidity affects first the more unsaturated fatty acids and, as it progresses, polyunsaturated fatty acids polymerise or break down to smaller molecules with fewer double bonds. IVs tend to fall, as does the proportion of unsaturated fatty acids to total fatty acids. This can be illustrated by considering the changes that take place in artificially autoxidised groundnut oil, heated for 20 h at 100 °C (Table 3.5). The most noticeable changes in the course of holding the oil, exposed to air, at 100 °C for 20 h, are the fall in IV and the loss of original monomeric fatty acids. There is no marked trend in the proportions of surviving individual fatty acids.

It must be stressed that the above conditions represent a drastically accelerated oxidation situation not likely to be of interest in practice, and that changes in IV or actual losses of fatty acids will only be marginal under practical conditions.

3.6 Peroxide values

Peroxide values (PV) are very sensitive indicators of the early stages of oxidative deterioration. Though peroxides and hydroperoxides of oils and

Table 3.5 Changes in iodine value and component fatty acids in groundnut oil heated at 100 °C for 20 h

	Original oil	Heated oil
IV	109.8	94.7
Component fatty acids (%):		
14:0–24:0	19.6	16.6
16:1–22:1	60.3	54.7
18:2	20.1	17.0
Total	100.0	88.3

fats are themselves tasteless, their presence is a sure indicator of flavour deterioration to come, since they are comparatively unstable. They will inevitably break down at ambient temperatures to yield a range of off-flavours deriving from small molecules, especially carbonyl compounds. At 100 °C breakdown takes place in a matter of days, and at 200 °C it can be complete in an hour or two. Thus, in frying at 180 °C, a partially deteriorated oil can exhibit a fall in PV due to decomposition of peroxides. Likewise in the final step of oil refining, deodorisation, at 220 °C, an important objective is the reduction of PVs to zero.

Hence, whilst PV is a most valuable indicator of developing rancidity in an oil stored at ambient or low temperatures, the thermal history of the oil needs to be known if one is to use PV, for example, for the assessment of frying oils. At sufficiently elevated temperatures, whilst oxygen is steadily being absorbed, peroxides are decomposing more rapidly than they are being generated. Such oils can have poor flavours and reduced IPs, despite having comparatively low PVs.

3.7 Newer methods based on oxidation intermediates

PV provides a means for predicting the risk of the development of flavour rancidity, but alternative procedures based on the further transformation of hydroperoxides are possible. Figure 3.2 indicates two of these.

Conjugated hydroperoxides from linoleic acid can dehydrate to form oxodienes, characterised by a UV absorption maximum at 275 nm, which is the basis for the quantitative estimation of oxodiene value (OV). Alternatively, both conjugated hydroperoxides and oxodienes can together be reduced by sodium borohydride to the corresponding alcohols, which are then dehydrated to form conjugated trienes (from linoleic acid) or conjugated tetraenes (from linolenic acid). The conjugable oxidation

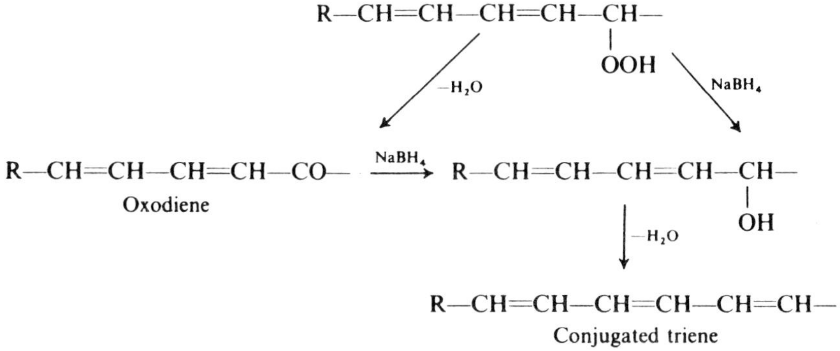

Figure 3.2 Transformation products of conjugated hydroperoxides.

products value (COP) is a combination of absorbance data at two wavelength maxima, 268 nm (trienes) and 301 nm (tetraenes).

3.8 Methods based on physical measurements

3.8.1 Thermogravimetric analysis

It has long been known that the total lipids increase in mass during the early stages of oxidative deterioration due to the conversion of fatty acids to hydroperoxides. Thermogravimetric analysis is the instrumental method for monitoring changes in mass, and this has been used successfully by some groups for the determination of IPs in heated oils. Thus soyabean oil had an IP of 25–31 min at 150 °C, although the IP increased to 1000 min at 100 °C.[12]

3.8.2 Chemiluminescence

Weak chemiluminescence occurs during the oxidative deterioration of edible oils and fats. This is due at least in part to the decomposition of hydroperoxides. Chemiluminescence has been found to correlate well with oxidised flavour in milk powder and reconstituted milk[13] but although chemiluminescence increased during the oxidation of milk fat after an initial fall, it decreased during the oxidation of soyabean and sunflower oils[14] and is therefore unlikely to be widely applicable. Enhancement of chemiluminescence by the effect of light or chemical reagents, e.g. sodium hypochlorite, allows lipid hydroperoxides to be detected at low levels.[15]

One problem with chemiluminescence is that its intensity is strongly dependent on hydroperoxide structure, with linolenic acid hydroperoxide emitting light ten times more strongly than linoleic acid hydroperoxide.[15]

3.8.3 Other physical measurements

Infrared spectroscopy has been used to monitor oxidative deterioration of an oil by measurement of the intensity of the band at 3450 cm^{-1}, which is due to the hydroxyl stretching of a hydroperoxide.[16] However, the absorbance is weak and the method is unlikely to have sufficient sensitivity to be developed into a general method.

Analysis of hydroperoxides by high performance liquid chromatography (HPLC) is becoming more common for studies of the mechanism or product composition arising from lipid oxidation. However, the technique has not been applied as a routine method for monitoring rancidity of oils.

Differential scanning calorimetry (DSC) has been employed as a rapid method for assessing the oxidative stability of oils.[17] This method relies on the heat change when unsaturated lipids react with oxygen, and it was found that IPs assessed in their way are much shorter than with the active oxygen method (AOM) method.

Ultraviolet absorption by conjugated dienes at 232–234 nm or conjugated trienes at 268 nm has been used to monitor oxidative deterioration of lipids as in IUPAC method 2.505. However, oxidation products formed from antioxidants during the induction period of an oil may interfere with this determination. For example, α-tocoquinone formed from α-tocopherol has peaks at 267.3 and 259.2 nm, and the UV absorption from this compound contributes to absorption at 232 nm.

3.9 Correlations between methods for evaluating rancidity

Working with carefully controlled model systems, subjected to conditions promoting accelerated deterioration, it is possible to determine the extent to which the various methods agree with each other. This is normally done either by regression analysis or by calculating correlation coefficients. The former is not possible in the present case as it is not feasible to keep one factor constant whilst altering others, as in the analysis of variance. In the present situation one is considering the relationships between two random variables, so the calculation of correlation coefficients is appropriate.

The correlation coefficient (r) between two variables (x and y) is given by:

$$r = \frac{1}{N}\sum\left(\frac{x - \bar{x}}{\sigma^x}\right)\left(\frac{y - \bar{y}}{\sigma^y}\right)$$

where $\bar{x}$ and $\bar{y}$ are mean values of x and y, σ^x and σ^y are standard deviations of x and y, and N is the number of observations.

In our studies,[2] accelerated deterioration was achieved by heating refined groundnut oil at 100 °C for 20 h, withdrawing samples every 2 h for test. For the earlier stages of deterioration (0–10 h) correlation coefficients were as shown in Table 3.6. In this early stage FS correlated well with IP, IV and AV, but less well with PV. This is not surprising because flavour deterioration can be slight, despite the formation of considerable amounts of tasteless peroxides.

When the heating period was longer (0–20 h) a second set of correlation coefficients (Table 3.7) was obtained. Again, FS correlated well with IP and IV, but the relationship between FS and AV had become more complex and correlation was poor. Correlations of FS with PV and OV

Table 3.6 Correlations between rancidity data. (Refined groundnut oil at 100 °C, 0–10 h)

	PV	IP	AV	IV	COP	OV	FS
PV	1.00						
IP	−0.96	1.00					
AV	0.85	−0.92	1.00				
IV	−0.82	0.94	−0.95	1.00			
COP	0.97	−0.95	0.85	−0.81	1.00		
OV	0.89	−0.93	0.97	−0.95	0.89	1.00	
FS	0.87	−0.96	0.99	−0.99	0.91	0.95	1.00

Table 3.7 Correlations between rancidity data. (Refined groundnut oil at 100 °C, 0–20 h)

	PV	IP	AV	IV	COP	OV	FS
PV	1.00						
IP	−0.95	1.00					
AV	0.67	−0.66	1.00				
IV	−0.96	0.96	−0.80	1.00			
COP	0.73	−0.81	0.48	−0.77	1.00		
OV	0.87	−0.97	0.76	−0.92	0.63	1.00	
FS	0.92	−0.98	0.73	−0.97	0.76	0.98	1.00

were, however, now much better, reflecting the validity of using oxidation intermediates as criteria of flavour determination in more advanced stages.

3.10 Accelerated test strategy

A major drawback of most current tests designed to evaluate stability or shelf life is that because of the need to obtain results expeditiously, autoxidation has to be artificially accelerated. This may be achieved by heat, irradiation, metal catalysis, oxygenation or enzyme action. In practice it normally involves heating the sample under controlled high temperature conditions. It is already accepted that irradiation, metal catalysis and oxygenases promote specific autoxidation reactions and have been used to monitor autoxidation under relevant conditions. However, practical situations also involve spoilage through autoxidation, e.g. the bulk, long term storage of oils in glass- or polythene-lined drums, sealed under 'oxygen-free' nitrogen. It is not sufficiently appreciated that evaluation of stability by accelerated heating tests is just as irrelevant in such cases as other accelerated tests, because mechanisms can be different from those encountered in practical situations.

The autoxidation of propyl linoleate has been shown, from its profile of volatile end-products, to take a significantly different course at 180 °C as compared with 70 °C.[18] Similarly, contrasting data have been obtained with ethyl linolenate at 250 °C as compared with 70 °C.[19] In studying the effects of a synergist on the protective properties of primary antioxidants it has been shown that stabilisation varies sharply with test (Rancimat) temperature.[20] Such studies suggest that, whilst conventional accelerated tests, such as the Swift test, may be realistic in the evaluation of oils and fats required to withstand high temperatures, as in frying or baking, they are misleading if extrapolated to ambient storage conditions.

Increasing understanding of the physiological and nutritional properties of polyunsaturated oils is leading to the development of high 'PUFA' (polyunsaturated fatty acids) products by the food and pharmaceutical industries, especially the $n-3$ and $n-6$ series of fatty acids present in many vegetable oils and all oils of marine product origin.[21] In these, it is of paramount importance that changes in the chemical structures of the constituent fatty acids and the minor lipid components in the course of processing and use be minimised.[22,23] It is quite unrealistic for such products to attempt to evaluate stability by tests conducted at temperatures higher than any reached during processing or in use. An alternative low temperature, accelerated test is necessary.

An accelerated test under ambient conditions, in the dark and without the use of pro-oxidant additives, must depend on the provision of much greater access of oxygen than is available under normal bulk storage conditions. This can be achieved by exposure of the oil to atmospheric oxygen at a very high surface/volume ratio, e.g. as a shallow layer of oil in an open Petri dish. Peroxide values can be monitored as a function of time of exposure. The shelf life of polyunsaturated vegetables oils, especially those containing linolenic acid, can be predicted in trials lasting two or three weeks and that of marine oils within one week. For practical purposes quality assurance of sensitive oils depends usually on the retention of PVs below specified maximum values for specified minimum times. For example the purchaser may specify a PV of no more than 10 for at least three months at 5 °C. Predictive tests in such cases are of considerable value.

3.11 Conclusions

Oils and fats from different sources deteriorate at different rates and produce different, often distinctive, off-flavours. Even nominally the same oil will be subject to batch-to-batch variation, thereby reflecting not only its origin and the climatic conditions prevailing at the time of seed ripening, but also its post-harvest history, including methods used for expelling or extraction, and especially for refining, including conditions for post-refining storage. A major additional factor will be the effect of subsequent chemical processing, such as hydrogenation or interesterification, or physical processing, such as fractionation.

In the studies described in this chapter, only a few oils have been examined and the conclusions must be accepted with some reservations. Other common oils or fats may possibly show a different pattern of correlations, and this may also be true of blends of two or more raw materials. Further, most of the data referred to apply to dry systems and should not be assumed to be valid also for emulsions or more complex food systems.

Within these limitations, however, it has been established that off-flavour formation in deteriorating oils, as determined by sensory tests, does correlate very well negatively with induction period and iodine value. Other criteria, such as peroxide value and anisidine value, though useful, do not correlate so precisely with the flavour score. Since sensory tests are, in practice, difficult to establish routinely, though sometimes essential, the analyst concerned with stability assessment would be well advised to supplement them with such chemical and, in particular, with automated induction period measurements. However, accelerated storage tests involving high temperatures can be misleading. It is hoped that an accelerated test will be eventually calibrated against and standardised for evaluation of stability at ambient temperatures.

References

1. Forss, D. A. (1972). Odour and flavour compounds from lipids. In *Progress in the Chemistry of Fats and Other Lipids*, Vol. 13, (ed. R. T. Holman), Pergamon Press, Oxford, pp. 177–258.
2. Odumosu, O. T., Sinha, J. and Hudson, B. J. F. (1979). Comparison of chemical and sensory methods of evaluating thermally oxidised groundnut oil. *J. Sci. Food Agric.*, **30**, 515–20.
3. AOCS Recommended Practice Cg 1–83. (1987). *Official Methods and Recommended Practices of the American Oil Chemists Society*, 3rd edn, Champaign, Illinois, USA.
4. Evans, C. D., List, G. R., Hoffmann, R. L. and Moser, H. A. (1969). Edible oil quality as measured by thermal release of pentane. *J. Am. Oil Chem. Soc.*, **46**, 501–4.
5. Gordon, M. H. (1987). Headspace gas analysis by gas chromatography. *Int. Analyst*, **1**, 16–20.
6. Frankel, E. N. and Tappel, A. L. (1991). Headspace gas chromatography of volatile lipid peroxidation products from human red blood cell membranes. *Lipids*, **26** (6), 479–84.
7. Snyder, J. M. and Mount, T. L. (1990). Analysis of vegetable oil volatiles by multiple headspace extraction. *J. Am. Oil Chem. Soc.*, **67**, (11), 800–803.
8. Snyder, J. M., Frankel, E. N. and Warner, K. (1985). Headspace volatile analysis to evaluate oxidative and thermal stability of soybean oil. *J. Am. Oil Chem. Soc.*, **63**, 1055–8.
9. Kirk, R. S. and Sawyer, R. (1991). *Pearson's Composition and Analysis of Foods*, 9th edn, Longman, Harlow.

10. Hudson, B. J. F. and Ghavami, M. (1984). Phospholipids as antioxidant synergists for tocopherols in the autoxidation of edible oils. *Lebensm-Wiss. u. Technol.*, **17**, 191–4.
11. Fishwick, M. J. and Swoboda, P. A. T. (1977). Measurement of oxidation of polyunsaturated fatty acids by the spectrophotometric assay of conjugated derivatives, *J. Sci. Food Agric.*, **28**, 387–93.
12. Mikula, M. and Khayat, A. (1985) Reaction conditions for measuring oxidative stability of oils by thermogravimetric analysis. *J. Am. Oil Chem. Soc.*, **62**, 1694–8.
13. Timms, R. E., Roupas, P. and Rogers, W. P. (1982) Determination of oxidative deterioration of milk powder and reconstituted milk by measurement of chemiluminescence. *J. Dairy Res.*, **49**, 645–54.
14. Timms, R. E. and Roupas, P. (1982). The application of chemiluminescence to the study of the oxidation of oils and fats. *Lebensm-Wiss. u-Technol.*, **15**, 372–7.
15. Yamamoto, Y., Niki, E., Tanimura R. and Kamiya, Y. (1985). Study of oxidation by chemiluminescence IV. Detection of low levels of lipid hydroperoxides by chemiluminescence. *J. Am. Oil Chem. Soc.*, **62**, 1248–50.
16. Takasago, M., Horikawa, K. and Masuyama, S. (1979). *J. Jpn. Oil. Chem. Soc. (Yukagaku)*, **28**, 291–4.
17. Cross, C. K. (1970). Oil stability. A DSC Alternative for the Active Oxygen Method, *J. Am. Oil Chem. Soc.*, **64**, 993–6.
18. Henderson, S. K., Witchwoot, A. and Nawar, W. (1980). The autoxidation of linoleates at elevated temperatures. *J. Am. Oil Chem. Soc.*, **51**, 409–13.
19. Lomanno, S. S. and Nawar, W. (1982). Effect of heating temperature and time on the volatile oxidative decomposition of linolenate. *J. Food Sci.*, **47**, 744–6 and 752.
20. Dziedzic, S. Z. and Hudson, B. J. F. (1984). Phosphatidyl ethanolamine as a synergist for primary antioxidants in edible oils. *J. Am. Oil Chem. Soc.*, **61**, 1042–5.
21. Hudson, B. J. F. (1987). Oilseeds as sources of essential fatty acids. *Human Nutr: Food Sc. Nutr.*, **41F**, 1–13.
22. Jawad, I. M., Kochhar, S. P. and Hudson, B. J. F. (1983). The physical refining of edible oils 1. *Lebensm-Wiss. u. Technol.*, **17**, 289–93.
23. Jawad, I. M., Kochhar, S. P. and Hudson, B. J. F. (1984). The physical refining of edible oils 2. *Lebensm-Wiss. u. Technol.*, **17**, 155–9.

Further reading

Simic, M. G. and Karel, M. (eds) (1979). *Autoxidation in Food and Biological Systems*, Plenum Press, New York; especially the following chapters:
  Frankel, E. N., Analytical methods used in the study of autoxidation processes, pp. 141–70.
  St Angelo, A. J., Legendre, M. G. and Dupuy, H. P., Rapid instrumental analysis of lipid oxidation products, pp. 171–83.
  Matsushita, S. and Asakawa, T., Simplified tests for fat deterioration, pp. 185–90.
  Nawar, W. W. and Witchwoot, A., Autoxidation of fats and oils at elevated temperatures, pp. 207–21.
Meara, M. L. (1980). Problems of fats in the food industry. In *Fats and Oils: Chemistry and Technology* (eds R. J. Hamilton and A. Bhati), Applied Science Publishers, London, pp. 193–214.
Supran, M. K. (ed.) (1978). *Lipids as a Source of Flavour*, ACS Symposium 75, American Chemical Society, Washington, DC.
Chan, H. W.-S. (ed.) (1987). *Autoxidation of Unsaturated Lipids*, Academic Press, London, New York, especially Pokorny, J., Major factors affecting the autoxidation of lipids, pp. 141–206.
Padley, F. B. and Podmore, J. (eds) (1985). *The Role of Fats in Human Nutrition*, Ellis Horwood, Chichester.
Hudson, B. J. F. (ed.) (1990). *Food Antioxidants*, Elsevier Applied Science, London and New York.

4 Practical measures to minimise rancidity in processing and storage
K. G. BERGER

4.1 Introduction

The term 'rancidity' is quite properly applied to fat that has deteriorated due either to oxidation or to hydrolysis. This chapter deals only with the effects of oxidation and their prevention.

In 1791 the distinguished French chemist Chaptal wrote the following: 'Oil easily combines with oxygene. This combination is either slow or rapid. In the first case rancidity is the consequence, in the second inflammation'. Fortunately food oils normally fall into Chaptal's 'slow' reaction category, and we do not often have to reach for the fire extinguisher!

Other chapters deal fully with the chemistry of oxidation of fats. Here we need only briefly remind ourselves of the four important influences on the rate of reaction:

1. It is obvious that oxidation cannot take place in the absence of oxygen (or air). While such a state is often difficult to achieve, a great deal can be done to minimise contact with air.
2. The rate of reaction of oxygen with fats roughly doubles for every 10 °C increase in temperature.
3. Some metals act as pro-oxidant catalysts in small amounts. The effect of 0.02 ppm copper or 0.5–1 ppm iron can be readily demonstrated. Traces of oxidised fat also act as catalysts and therefore cleanliness of plant is of extreme importance in the minimisation of oxidation.
4. Light promotes oxidation, initially through a singlet oxygen reaction which is not inhibited by antioxidants. Shorter wavelengths and higher intensities are more effective.

Two other general points may be made. The more unsaturated oils are more sensitive to oxidation, the reactivity of the constituent fatty acids being in the order oleic ≪ linoleic ≪ linolenic and more unsaturated ones. Secondly vegetable oils are protected from oxidation to varying extents by their content of natural antioxidants, mainly the tocopherol group (Vitamin E).

As a rough guide, the commonly used food oils and fats can be listed in a ranking order.

- Cocoa butter   Most resistant
- Coconut oil
- Palm kernel oil
- Palm oil
- Most hydrogenated oils
- Groundnut (peanut) oil
- Sesame oil
- Cottonseed oil
- Olive oil
- Maize oil
- Sunflower oil
- Soya bean oil
- Rapeseed oil
- Butterfat
- Beef fat
- Lard   Least resistant

Such a list should be used with caution. It applies only if the oil has not been abused beforehand. In simplistic terms, an edible oil, like the proverbial cat, can be thought to start with nine lives. Inevitably storage, handling and processing result in some loss of 'life'. Quality will be optimised if we can ensure that the ultimate consumer gets the oil (perhaps in a food product) with as many 'lives' intact as possible. What follows is intended to provide some guidelines to this objective.

4.2 Design factors

The design of any plant is a matter of compromise. However, the oils and fats technologist must argue on behalf of the materials for which he is responsible, because the engineers and draughtsmen involved in the project will not understand their peculiar behaviour. One requirement over which there can be no compromise is the absolute exclusion of copper or any copper-containing alloy from contact with oils and fats.

4.2.1 Storage tanks and process vessels

Process vessels should be of stainless steel. Although it may be argued that mild steel is good enough when only refined oils are being handled, it is not inert and it is not easy to clean or to prevent from rusting. It allows no 'leeway' for the odd occasion when a process hitch occurs and material has to be held at high temperature, or for a prolonged period. One ton of fat makes about half a million pieces of biscuit or portions of ice cream. An off-flavour can lose many customers.

Storage tanks of stainless steel are also recommended, but if large they may be too costly. In that case one possibility is a mild steel tank with a thin sheath of stainless steel. Alternatively mild steel tanks may be coated with an inert epoxy or polyurethane resin. Such coatings must be applied by specialists on a newly sand-blasted surface and given sufficient time to cure. The coatings have limited resistance to abrasion and to high temperature. Steam coils should be no closer than 0.4 m to the coatings. In the author's experience, with care, epoxy coatings on static tanks are very satisfactory for 10–20 years. One of their advantages is their ease of cleaning.

The shape of tanks is often determined by space available and the permissible floor loading. For a given volume tall narrow tanks minimise the surface of oil exposed to air.

A simple means of further reducing the surface area of oil exposed to the air during storage is to fit a floating lid to the storage tank. To illustrate the effectiveness of this, Table 4.1[1] summarises the results of a test in which 500 tons of crude palm oil were divided between two tanks, one being fitted with a floating lid. Storage was at 50 °C.

Tanks should be designed to drain completely. Suitable alternative shapes for the bottom are conical, dished or sloping towards one side. In each case the outlet should be at the lowest point. Where an oil is being stored which throws a deposit or 'foots', for example a crude oil, it is an advantage to have a second outlet about 1 ft (30 cm) above the bottom, so that if desired, it is possible to draw off clear oil and treat the sediment separately.

Table 4.1 Effect of lid on oil deterioration in a storage tank

Storage period (days)	Without lid		With lid	
	Peroxide value meq kg^{-1}	Total tocopherol	Peroxide value meq kg^{-1}	Total tocopherol
0	1.7	710	1.4	710
12	3.2	680	1.6	710
24	3.7	680	1.0	740
48	4.4	650	0.8	680
Ratio of exposed surface to weight of oil (sq. ft/ton)	30.7		0.36	

4.2.2 Tank heating

Process tanks usually require to be heated, and so will any storage tanks for oils that tend to solidify at room temeprature. The emphasis should be on good temperature control so that the oil is not overheated. For storage tanks, a hot water jacket or internal stainless steel coils through which hot water is circulated are preferred. The coils should be raised 8–30 cm above the base on short legs, and the hot water thermostatically controlled.

Large tanks are usually heated by coils using low pressure steam. A coil area of $0.05 \text{ m}^2 \text{ton}^{-1}$ capacity is advised for temperature maintenance, though less is adequate for a lagged tank. For melting, a higher capacity of $0.1 \text{ m}^2 \text{ton}^{-1}$ is required. When oil is stored for prolonged periods, it is allowed to cool, and some oils partially solidify. If the heating coils are completely covered by solid fat, there is a serious risk of local overheating, since convection currents will not be able to flow throughout the mass. Coils are therefore designed with a 'hairpin' up the side of the tank, which allows a passage to be melted through the solid crust.

A rather sophisticated heating system for large storage tanks is described by Kellens.[2] Four heaters are symmetrically disposed in the tank. Each heater consists of:

(a) a conically shaped steam coil covered by a metal cone open at both ends; and
(b) a series of four hollow cylindrical jackets mounted in line above the cone.

The conical coil and each jacket have separate steam supplies. Depending on the amount of oil required one or more of these heaters are activated. When switched on, each cone and its cylinders together act as a thermal pump which circulates the oil efficiently. The system is claimed to liquefy oil at minimum temperatures without any local overheating.

A possible alternative where a number of relatively small tanks are required, is to group them together in a heated room, provided the temperature required is uniform and still tolerable to operators.

Storage tanks situated in the open should always be lagged. This reduces heat loss, but it also reduces the effect of sunshine in summer. This can cause differential heating of the oil on one side only, giving rise to convection currents which mix the oil and increase the rate of oxidation.

In process tanks hot water circulation may not provide sufficiently fast heat-up, and steam coils may be required. Steam should always be low pressure, to minimise local overheating.

4.2.3 Pipelines

Pipelines should be of stainless steel, except for large storage installations, where this would be uneconomical. Alternatively it is now possible to have mild steel pipelines internally epoxy coated.

Pipelines should be laid to a fall, so that they drain completely to a valve placed at the lowest point. It is important to avoid U-bends in which oil can accumulate. Alternatively for long pipelines and large systems the pipelines should be designed for use with a 'pigging' system for line clearance. In this context a 'pig' typically consists of three or four dish-shaped plastic discs mounted centrally on a flexible rod. The discs fit the internal diameter of the pipeline. The pig is introduced through a side valve at the end of the pipeline, and forced through by compressed air. It pushes the contents out, and is removed through a second side valve at the far end. Any bends in the line must have a minimum diameter. Soft 'pigs' of a spongy plastic may also be used, but are less effective. The use of pigs is especially useful where different types of oil are handled in the same line. Valves in pipelines should not have any 'pockets' or corners in which residues can remain.

Pipelines used for filling tanks should preferably enter at the top of the tank and reach close to the bottom. There needs to be a siphon-breaker at the top. Such a pipe minimises aeration during pumping. Ships' tanks are often filled or emptied at a rate of more than 150 ton h^{-1} and, if the oil is allowed to cascade, there will be considerable contact with oxygen.

Heating for pipelines. Pipelines that are used for oils which solidify at ambient temperature require provision for heating. This may be by hot water jacket, electrical heating tape, resistance heating or by steam tracer lines. After use the pipeline should always be drained. If this draining is complete, it may be unnecessary to use the pipe heating in routine operation. Alternatively it may only be necessary to use heat during the start-up period. Depending on circumstances, the heating should be minimised, and in any case when the pipeline has been drained the heat must be turned off, as remaining thin films of oil will oxidise very quickly.

4.2.4 Pumps

Care must be taken that pumps do not draw air into the oil. They should not have phosphor bronze or other copper containing bushes.

4.2.5 Storage in drums

The effect of long term storage of oils in standard 40 gallon drums is illustrated by an experiment carried out on crude palm oil[3] (see Table 4.2). Storage was at Malaysian ambient temperature (25–30 °C). A

Table 4.2 Storage of crude palm oil bulk samples drawn from drums

Storage (days)	Peroxide value (Meq kg^{-1})		
	Uncoated		Coated
	No AO	Added AO	Added AO
300	13.5	1.9	0.8
500	25.4	3	1
600	2.6	3.3	1
730	24.4	3.6	1.6

AO = Antioxidant 110 ppm tertiary butylhydroquinone plus 365 ppm citric acid.

comparison was made between an epoxy-coated and an uncoated drum and between the presence and absence of antioxidant. The drums wre stored closed, but air was not rigidly excluded.

This test clearly shows that even in the presence of an effective antioxidant mixture, the inert coating gives added protection. Samples were also drawn from different levels of each drum after two years' storage. The analytical results are given in Table 4.3.

The palm oil was more than 80% liquid in the conditions of storage and oxidation had proceeded throughout the bulk of the oil. Similar effects were reported by Eapen.[4] Crude palm oil was stored in standard resin lined drums and polythene lined drums for 126 weeks. The analytical results are summarised in Table 4.4.

Table 4.3 Condition of crude palm oil after prolonged storage in drums. Peroxide value (meq kg^{-1}) and carotene (ppm)

Depth (in)	Control		Coated		Uncoated	
	PV	Carotene	PV	Carotene	PV	Carotene
6	47	269	2.05	651	4.8	600
12	48	269	2.25	671	4.7	608
18	29.6	299	1.84	680	4.4	613
24	19.4	444	1.74	642	4.4	595
30	–	–	1.38	634	3.2	590

Table 4.4 Peroxide value of crude palm oil during storage in drums

Storge (weeks)	0	16			126		
Type of lining	–	None	Resin	Polythene	None	Resin	Polythene
Iron (ppm)	7.6	35	–[b]	–[b]	35	–[b]	–[b]
Peroxide value[a]	Nil	+ve	Nil	Nil	4.8	0.69	1.9

[a]ml 0.002 N thiosulphate per g sample. [b]Virtually unchanged.

Heating for drums. Where solid fats have been stored or transported in drums, it may be necessary to melt the contents before use. A number of methods are commercially available.

1. Insertion of a live steam hose into the drum. This causes hydrolysis and local overheating. An additional process to separate the condensed water is needed.
2. Hot chamber heated by closed steam coils. This can be satisfactory if there is provision to drain off the fat as it melts, for example by inverting the drum over a drainage channel. However oil is liable to be contaminated by dirt.
3. Electrical heating by a hot plate. This is rather slow and may lead to local overheating.
4. Electrical heating by means of a hinged jacket that fits round the drum. This is better than method 3, above, but may lead to local overheating.
5. Electrical heating by induction. The heater consists of a metal sleeve which is lowered over the drum. A preferred system which uses less electricity than radiant elements, and does not give hot spots.
6. For open top drums a portable steam coil shaped in a horizontal spiral may be used. The tubing has an elliptical cross-section and cuts through the fat under its own weight. In the author's experience[5] fat with a melting point of 40–45 °C can be melted with this equipment without it reaching 50 °C.

4.2.6 Retail packaging

The factor of access of light comes into play here. An interesting series of storage tests was carried out by Becker and Niederstebruch.[6] They stored oils in bottles packed in air and under nitrogen, and stored either in the dark or in daylight. The flavour scores of the stored oils are summarised in Table 4.5.

The flavour scores were the averaged opinions of 12 tasters. A score of 7 or above is good, while below 5 is unpalatable. The results show clearly that residual oxygen even at low levels still caused deterioration, and in each case the effect of daylight caused significant acceleration.

While marketing considerations make it desirable for the customer to see the product, a transparent or translucent pack will enable surface oxidation to be promoted by light. For bottled oils some limited protection may be gained by the use of coloured glass or plastic, or by designing the pack so that the top part of the bottle, where the oil surface is, is protected by a sleeve or label, while the clarity and colour of the oil can still be seen lower down.

Table 4.5 Effect of oxygen and daylight on stored oils

Oil	Storage		Dissolved oxygen (mg/100 ml)	Flavour score after storage (days)				
				0	21	27	41	48
Cottonseed	Air	Dark	2.96	7.2	4.8	–	–	–
		Light	2.96	7.2	4.8	–	–	–
	Nitrogen	Dark	0.82	6.9	6.9	–	–	–
		Light	0.82	6.9	5.9	–	–	–
Soya	Air	Dark	2.52	7.2	–	–	–	4.5
		Light	2.52	7.2	–	–	–	3.6
	Nitrogen	Dark	0.66	7.1	–	–	–	5.9
		Light	0.66	7.1	–	–	–	5.1
Sunflower	Air	Dark	2.68	7.5	–	6.2	5.9	–
	Nitrogen	Dark	0.12	7.5	–	7.3	7.1	–

It is desirable to check the wrapping material for traces of pro-oxidant metals. A particular problem can arise with fried snackfoods having a relatively high oil content. Some of the oil is spread on the surface of the packing material and may oxidise rapidly. A point is reached when, on opening the packet, an unpleasant whiff of rancid smell is perceived, even though the contents themselves are still palatable. This thin film of rancid oil will soon promote rancidity in the product itself.

Rho and co-workers[7] reported the effect of various antioxidants either in the frying oil or coated onto the inner surface of the polythene packaging on the shelf life of instant noodles. They used organoleptic tests after storage at 63 °C. The results are summarised in Table 4.6.

The antioxidants used were 'Anoxomer', butylated hydroxyanisole (BHA) and butylated hydroxytoluene (BHT). Ethylenediamine tetra-acetic acid (EDTA) was also added in two tests to sequester traces of pro-oxidant metals. A positive effect was obtained.

The two experiments were carried out under different conditions and

Table 4.6 Days at 63 °C required to develop rancidity

Experiment I Antioxidant in oil		Experiment II Antioxidant on packaging	
Control	5	Control	18
200 ppm Poly 'A'	10	200 ppm TBHQ[a]	37
200 ppm BHA	11	500 ppm TBHQ[a]	40
200 ppm TBHQ	14	1 000 ppm TBHQ[a]	44
200 ppm TBHQ[a] 200 ppm EDTA	18		
200 ppm TBHQ[a] 500 ppm EDTA	20		

[a]Calculated relative to the oil content of the noodle (20%).

therefore the protective effects cannot be readily compared. It is clear however that coating of the packaging with antioxidant is an effective procedure. It must be done at an appropriate level, so that any antioxidant migrating into the food does not exceed the legally permitted amount.

4.3 Operational factors

4.3.1 Handling of bulk stock supplies

While an oil storage or process tank is in continuous use it will keep itself clean, provided no insoluble residues settle out. If, however, there is an appreciable period of interruption, then it should be assumed that any remaining oil has lost quality, and it should not be mixed with fresh oil. Storage tanks should always be completely emptied, inspected and cleaned if necessary before being refilled. Oils in stock should be used in proper rotation.

4.3.2 Effect of dissolved oxygen

A few investigators have attempted to quantify the effect on quality of oxygen dissolved in the oil. The solubility of oxygen in various oils was determined by Law and coworkers.[8] Results are summarised in Table 4.7. Measurements were made by the chemical method of McDowell[9] which also measures any peroxides present. A correction for peroxides was obtained from a measurement of dissolved oxygen content with a Beckmann Monitor II Dissolved Oxygen Analyser.

Heide-Jensen[10] measured the dissolved oxygen content of refined fish oil by means of a Clark oxygen electrode. He found that oxygen was absorbed at a rate of 0.1 mmol dm^{-2} of surface per 24 h at 20 °C. Samples of oil were stored until different proportions of oxygen were absorbed and were then refined and hydrogenated. Samples of the refined products were stored for 3 days at 50 °C either in air or in closed cans under inert gas, and they were tasted in pairs by a trained panel. It was established that oil which had absorbed 8 mmol oxygen/kg or more could be clearly

Table 4.7 Dissolved oxygen content of oils (ppm)

Temperature (°C)	30	40	45	50	60	70
Palm oil	46.0	–	39.4	–	39.7	28.8
Palm olein	47.7	51.2	45.5	36.4	36.0	32.4
Palm stearin	–	–	–	–	31.2	33.6
Corn oil	48.4	47.4	42.6	43.2	–	27.3

distinguished by taste. This level corresponded to an increase in peroxide value of 5.2 mmol kg^{-1}, but decomposition products of peroxides were not measured. The significance of this finding is that even mild peroxidation of the oil was not rectified by subsequent processing, but led to a lower end-product quality.

Becker and Niederstebruch[6] measured dissolved oxygen at various stages during the refinery process (see Table 4.8). The oil was held for 1 h at 70 °C in the intermediate tank, sufficient for a significant development of peroxides. Although these are destroyed in the deodoriser, their decomposition products can be expected to exert a pro-oxidant effect in the end-product, as demonstrated by Heide-Jensen.

Results more recently reported[11] on palm oil in two refineries show that present day process design still does not take oxygen contact sufficiently into account (see Table 4.9). The degumming and bleaching step reduces the oxygen content, but subsequent pumping increases it. Oxygen is removed in the deaerator, albeit not always effectively, and in the deodoriser, but readmitted during pumping into storage.

Table 4.8 Dissolved oxygen during processing

	Oxygen (mg/100 ml)	Peroxide value	Oxygen (mg/100 ml)	Peroxide value
Crude oil	0.16	5.8	1.04	12.9
Neutralised	0.33	5.7	0.45	11.2
Bleached	0	0	0	0
Last runnings from filter press	0.59	0	–	–
Intermediate holding tank	0.37	2.2	0.65	3.6
Deodorised oil	0	0	0	0

Table 4.9 Oxygen content of oil (ppm) during processing

Refinery	A Mean of 12 observations	B Single
Crude	15	32
Degummed, bleached	8	20
Filtered	–	8
In buffer tank[a]	26	24
From deaerator[b]	12	–
From heat exchanger	–	36
Fully refined	8	4
In storage tank[a]	31	24
After fractionation[a]	–	36

[a]Oxygen content increases each time oil is pumped into a tank. [b]Oxygen content after 'deaeration' varied from 0.3 to 30.6 ppm over 12 runs.

The dissolved oxygen content of several bulk parcels of oil was monitored during transport from Malaysia to Rotterdam.[12] The results together with peroxide values measured at the same time are summarised in Table 4.10. The dissolved oxygen content recorded is an average of measurements at various depths throughout the tank. These results present several points of interest. All the oxygen contents increased at loading, due to cascading into the tank. The initial oxygen content of the crude palm stearin was low, presumably because autoxidation was already proceeding. This continued during the voyage. The content of iron was rather high in this crude oil. The rates of oxidation of the other three oils were quite slow and consequently the dissolved oxygen content was not much changed.

These examples are sufficient to show that the exclusion of oxygen at any stage in the life of an oil has benefits in terms of the quality of the product in the hands of the ultimate consumer. The dissolved oxygen content of oils can be substantially reduced by sparging.[13] This involves the injection under pressure of an inert gas, preferably nitrogen with a minimal residual oxygen content, into the oil as it passes through a pipeline. The gas enters the oil stream in small bubbles through a diffuser mounted in line. Passage through a length of pipeline is then desirable to give time for equilibration. According to Henry's law, the solubility of oxygen in the oil depends on the partial pressure of oxygen in the atmosphere in contact with it. Therefore the oxygen partitions between the oil and the sparging gas. At equilibrium the oxygen content remaining in the oil is given by the equation:

$$f = \frac{1}{\frac{MR}{P_t} + 1}$$

Table 4.10 Dissolved oxygen content and peroxide values (meq kg^{-1}) during bulk transport

	Refined palm olein		Crude palm stearin		Refined palm stearin		Crude palm kernel oil	
	O_2 (ppm)	PV	O_2 (ppm)	PV	O_2 (ppm)	PV	O_2 (ppm)	PV
Before loading	30	0.8	7	4.0	26	0.8	43	0.4
After loading	39	0.8	26	4.5	35	0.9	47	0.4
After days voyage 21	39	0.8	1	5.2	24	0.3	34	0.7
23	39	1.6	1	5.7	24	1.0	36	1.0
25	35	1.6	1	5.9	27	1.7	30	1.4
At discharge	30	1.7	2	8.0	26	2.1	31	2.1
Iron content at loading (ppm)	0.4		13.5		1.8		1.1	

where f = ratio of remaining dissolved O_2 to original dissolved O_2, M = temperature dependent constant, R = ratio of volume nitrogen to liquid, P_t = pressure of system. A single-stage sparger gives 80–90% reduction in dissolved oxygen. Multiple-stage sparging is more effective.

Bek-Nielsen[14] was able to obtain significant improvements in the arrival quality of crude palm oil shipments from Malaysia to Europe by sparging the oil with inert gas. Results obtained in normal and protected shipments are shown in Table 4.11.

In the last 10 years the amounts of edible oils transported in bulk have increased enormously and now total more than 20 million tons worldwide. While the shipments monitored in Table 4.10 showed only a slight increase in oxidation, this is by no means always the case. Shipments of fully refined oils from Malaysia to South Korea monitored during 1983/4 had peroxide values of 5–9 meq kg^{-1} on discharge.

The value of nitrogen protection during bulk shipment of refined oil was recently reported by Elias and Zainal Sham.[15] Table 4.12 shows the peroxide values of untreated and nitrogen-sparged palm olein. Nitrogen sparging was subsequently also provided for the treated oils when successive lots were drawn from the storage tank and despatched. The results are given in Table 4.13. The protective effect of the nitrogen treatment is clear.

Similar results have been reported for road shipments across Australia.[16] Fully refined vegetable oils are routinely shipped by road tanker from Sydney to Perth, a 9-day journey. The peroxide value on arrival is

Table 4.11 Oxidation of crude palm oil during transport

Sample	Peroxide Value (meq kg^{-1})	Anisidine value
Normal range at production	0.2–2.0	1.5–2.5
Normal range at loadings in Malaysia	2.0–6.0	2.5–4.0
Normal range at arrival in Europe	6.0–12.0	4.0–6.0
Sparged consignment at loading	1.9	1.8
at discharge	2.5	4.7

Table 4.12 Nitrogen sparging of bulk shipments of refined palm olein

	Sparged PV (meq kg^{-1})	Untreated PV (meq kg^{-1})	Sparged PV (meq kg^{-1})	Untreated PV (meq kg^{-1})
Loading	0.4	0.69	0.50	0.48
Arrival	0.45	2.35	0.68	3.10
After 43 days storage	0.8	5.0	1.2	5.5

Table 4.13 Nitrogen sparging in road and rail transport

Transport	N$_2$ treated PV (meq kg^{-1})	Untreated PV (meq kg^{-1})
Road (1 day)		
Loading	0.6–1.0	2.4–4.5
Arrival	0.6–1.5	3.7–6.0
Rail (7–14 days)		
Loading	0.5–0.8	2.2–3.7
Arrival	0.8–2.0	4.9–9.0

below 1 and the oils do not require any further processing before use. The sparging process consisted of injecting nitrogen under pressure into the oil as it was pumped into the tanker. Some nitrogen dissolves in the oil and is released when the oil comes to equilibrium in the road tanker; it then forms an inert blanket.

Becker and Niederstebruch[6] also investigated the advantages of nitrogen protection in the final stages of margarine processing, texturisation in a Votator. Margarines processed in air and under nitrogen were stored at 18° C in the dark and tasted at intervals by a trained panel. The results of two separate experiments are summarised in Table 4.14. Flavour scores below 5 are unacceptable. The extent of oxygen reduction achieved varies in the two experiments, but shelf life was significantly increased in both.

The costs of nitrogen protection can be offset against the higher costs of processing a partly oxidised oil. The benefits of improved stability in the consumers' hands are less easily quantified, but will assume increased importance as regulations on date marking of foods become stricter.

Table 4.14 Nitrogen versus air in margarine processing

Sample and gas compostion	Flavour score after storage (days)					
	0	24	29	37	81	88
In air O$_2$ 62 mg/100 g N$_2$ 6.15 mg/100 g	7.1	–	5.5	–	4.6	–
In nitrogen O$_2$ 1.21 mg/100 mg N$_2$ 7.96 mg/100 g	7.0	–	7.0	–	6.5	–
In air O$_2$ 2.42 mg/100 g N$_2$ 6.82 mg/100 mg	6.9	5.8	–	5.7	–	4.8
In nitrogen O$_2$ 0.13 mg/100 g N$_2$ 8.52 mg/100 g	7.0	6.8	–	6.5	–	5.9

4.3.3 Temperature control

The use of elevated temperatures is necessary at various stages in the refining process and in most other food processes. The practical effect of a higher temperature is seen in the following chart (Figure 4.1). In a laboratory experiment 400 g samples of commercial shortening blends were stored in ovens at 50 and 60 °C, respectively. The figure gives the

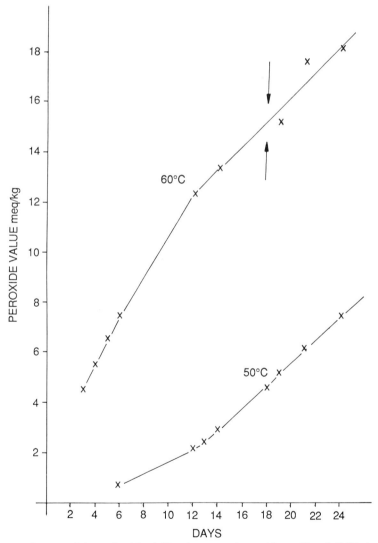

Figure 4.1 Storage of shortening blend. Development of peroxides at 50 and 60 °C. Arrows indicate point of perceptible off-flavour.

Table 4.15 Storage temperature[17]

Products	Minimum (°C)	Maximum (°C)
Palm oil (refined or crude)	50	55
Palm stearin (refined or crude)	55–60[a]	65–70[a]
Palm olein (refined or crude)	30	35
Palm oil mid-fraction	40	45
Palm kernel and coconut oil	30	35
Palm kernel olein	30	35
Palm kernel stearin	40	45
Tallow	55	60
Fish oil	25	30
Liquid vegetable oils (e.g. soyabean, rapeseed)	20	25
Hardened oils	10 above slip point	15 above slip point

[a]The lower temperatures apply to soft grades, while the higher temperatures are necessary for hard grades. The temperatures apply to both crude and processed oils in each grade.

results of peroxide value determinations at intervals. Tasting tests showed distinct off-flavours after 18 days' storage at 60 °C as indicated by the arrow, while the blend stored at 50 °C remained bland throughout. The following general advice is offered:

1. Temperatures should be automatically controlled and recorded whenever possible.
2. Processes such as bleaching in the refinery should be optimised towards the use of lower temperatures where possible.
3. When a process is completed, the oil should be cooled in a heat exchanger as soon as possible to the temperature appropriate for the next process step.

If oils are to be stored for a long time, they should be allowed to get cold. For storage of bulk stocks awaiting use, storage time should be minimised and maximum temperatures should be observed according to Table 4.15.[17]

Acknowledgements

Thanks are due to Mr Hyam Myers and Mr A. F. Mogerly for providing information and for helpful discussion.

References

1. Harrisons & Crosfield Ltd, (1973). *Internal Research Report*.
2. Kellens, E. (1972). *Oleagineux*, **27**(3) 157–60.

3. Chong, C. L. (1985). *PORIM Annual Research Report*, Vol. 2, p. 277
4. Eapen, P. I. (1977). *Technical Consultation on Oil Crops for West and Central Africa*, FAO, Benin City, Nigeria.
5. Leong, W. L. and Berger, K. G. (1982). Storage, handling and transportation of palm oil products, *PORIM Technology*, No. 7, Palm Oil Research Institute of Malaysia, Kuala Lumpur.
6. Becker, E. and Niederstebruch, A. (1966). *Fette Seifen Anstrichm.*, **68**(2) 1–12.
7. Rho, K. L., Selb, P. A., Chung, O. K. and Chung, D. S. (1986). *J. Am. Oil Chem. Soc.*, **63**, 251.
8. Law, K. S., Timms R. E. and Wong. (1984). *PORIM Research Report*, PO (82) 84.
9. McDowell, A. K. R. (1963). *J. Dairy Res.*, **30**, 399.
10. Heide-Jensen, J. (1965). *Proceedings of 4th Scandinavian Symposium of Fats and Oils*, ABO, Gordon & Breach, New York, pp. 131–40.
11. Berger, K. G. (1986). *J. Am. Oil Chem. Soc.*, **63**(2) 217–22.
12. Elias, B. A. (1985). *PORIM Research Report*, TAS/19–1/84.
13. British Oxygen Co. Ltd, Data Sheet—Nitrogen Processes No. 4.
14. Bek-Nielsen, B. (1977). *Oleagineux*, **32**(10) 437–41.
15. Elias, B. A. and Zainal Sham (1988). *Proceedings of International Symposium on Palm Oil*, Kuala Lumpur, Palm Oil Research Institute of Malaysia, p. 235.
16. Smith, R. (1984). Private communication.
17. Codex Alimentarius Commission—Oils and Fats Committee. Draft International Code of Practice for Storage and Transport of Edible Oils & Fats in Bulk. Document Alinorm 87/17.

5 The use of antioxidants
P. P. COPPEN

5.1 Basic principles and definitions

Antioxidants are often grouped together with the larger group of food additives that extend shelf life, the preservatives, and it is useful to start with a definition. Food additive law provides one: In the UK the use of these products is controlled by the Antioxidant in Food Regulations 1978 which give a legal definition:

> '... any substance which is capable of delaying, retarding or preventing the development in food of rancidity or other flavour deterioration due to oxidation...'

Oxidation of food components may occur in the aqueous or lipid phase. This chapter is mainly concerned with oxidation of lipids in foods. The fat or oil content of a wide range of foods is prone to oxidation, as a look at the list of permitted applications in these regulations makes clear. The structural chemical principles and probable reaction mechanisms of antioxidant activity are covered in other chapters, as is the measurement of rancidity. For our purposes it is only necessary to keep certain key points in mind:

1. It is the vicinity of the carbon–carbon double bond that is attacked by oxygen, termed 'autoxidation' as it occurs without our intervention.
2. Oils with a high proportion of these 'unsaturated' bonds in their triglyceride fatty acids are more prone to 'autoxidation'; for example, soyabean oil is much less stable than coconut oil.
3. The breakdown products of this oxidation—ketones, aldehydes and low molecular weight fatty acids—are relatively volatile compounds and give rise to the characteristic off-flavours and odours associated with oxidative rancidity.
4. Some of these compounds have very low odour/flavour thresholds, and the presence of very small amounts of these—levels as low as parts per million or parts per billion—can make a food 'rancid'.

Perhaps it should be stressed early on in any discussion of the use of antioxidants that, notwithstanding the wording of the UK legal definition, for all practical purposes none of the available antioxidants entirely prevent autoxidation—they delay it. Delaying the onset of rancidity means that the foodstuff is acceptable for a longer period; it has a better shelf life. The amount of this extension of shelf life will vary considerably depending of the composition and processing of the food, the antioxidant, packaging and storage conditions. In extreme cases the shelf life may be so long that it appears that rancidity has been prevented.

5.2 Why use an antioxidant?

Antioxidants are only one means of fending off oxidation; there are others, e.g. vacuum or controlled atmosphere packaging or packing under an inert gas to exclude oxygen or/and refrigeration/freezing, both of which greatly reduce the rate of autoxidation. Unfortunately, these are not always applicable. Furthermore, it is seldom realised how little oxygen is needed to initiate and maintain the oxidative process, or how difficult and expensive in equipment terms it can be to remove the last traces of air from a food product. Consequently it is reasonable to combine the use of antioxidants with, say, inert gas packing. Using an antioxidant should be seen as one of several measures available but, used properly, it is generally effective, easily applied and inexpensive. (As an illustration of the order of cost involved, adding an antioxidant might increase the cost by less than half a percent.) The prime justification for using an antioxidant is one of need: an antioxidant can extend the shelf life of a food, reducing wastage and complaints; it can reduce nutritional losses (oil-soluble vitamins, e.g. vitamin A, are prone to oxidation); and a very important point for the food technologist, it can widen the range of fats which can be used. Using an antioxidant enables the food manufacturer to smooth out differences in the stability of fats/oils and renders the food product less specific in terms of ingredient requirements. This offers more scope for cost control without jeopardising the product quality or shelf life. Without an effective antioxidant, lard, for example, would find far fewer uses.

5.3 Requirements of an ideal antioxidant

An ideal antioxidant meets the following demands:
1. Safe in use,
2. Should impart no odour, flavour or colour,
3. Effective at low concentrations,
4. Should be easy to incorporate,

5. Should survive cooking processes such as baking and frying,
6. Should be available at a low cost-in-use.

This list is stringent and the number of antioxidants that meet the demands is limited. Those currently used are discussed later.

Although safety in use is paramount it is becoming increasingly clear to scientists and regulatory authorities throughout the world that when considering safety we have to decide whether the degree of risk is acceptable. Nothing can be said to be without risk; food is no exception. In the limited field of foods and antioxidants we know less about the hazards of oxidised fats and oils than about the safety of the antioxidants. The picture is complicated as they display paradoxical behaviour, for example showing both anti-carcinogenic and carcinogenic effects in differing experimental protocols. Effects may appear dose related and parallel the pro- and anti-oxidant effects we see in food systems.[4] However, despite occasional media claims based on anecdotal evidence, we can say that all the currently permitted antioxidants have undergone extensive safety evaluation and have survived scientific scrutiny throughout almost two generations of widespread use in food.

Point 5 is very important as large quantities of fats are used in food manufacturing processes where high temperatures are encountered and a good shelf life is desired of the final product.

High temperature is likely to damage the fat and reduce any natural stability. In many cases this is made doubly difficult by the fact that the final product, a biscuit for example, has a fine crumb structure, which means that there is a large surface area exposed to air, encouraging autoxidation. This is the reason why the bulk of antioxidant usage appears in manufactured foodstuffs, particularly biscuits, snack foods and noodles. The ability of an antioxidant to survive the high temperatures encountered in cooking processes is often referred to as its 'carry-through' property.

5.4 Oxidation of lard

Figure 5.1 illustrates how the oxidation of a typical animal fat, in this case lard, proceeds against time. The peroxide value has been used as an indication of the degree of oxidation, and it has been found that it correlates fairly well, although not always exactly, with flavour. In fact by measuring peroxide value we are monitoring the precursors of products that affect the flavour. As can be seen, the oxidation process divides into two sections. The first section where the slope is moderate is often referred to as the initiation or induction period. It is generally accepted that oxidation proceeds via a free radical mechanism and during the

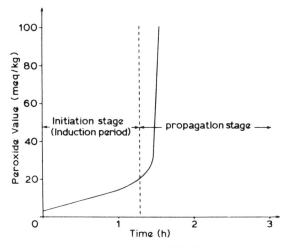

Figure 5.1 Oxidation against time for lard at 100 °C by the active oxygen method.[1]

initiation period, energy in the form of heat, light, and so on, reacts with the fat to produce free radicals. At a certain stage the concentration of free radicals reaches a critical level and a rapid self-propagating mechanism takes over. This is referred to as the 'propagation stage'. During this stage there is a very rapid increase in peroxide value. Figure 5.1 is an ideal example, somewhat stylised but generally true for animal fats. Vegetable oils do not usually show such a clear-cut change of slope.

The peroxide value can be misleading, and is only one of the indicators monitored (see chapter 2). It is possible for a rancid sample to have a low peroxide value. If the peroxide value is observed for long enough it will be seen to reach a maximum and then finally decline. This occurs when the rate of decomposition of peroxides or hydroperoxides into volatile compounds outstrips their rate of formation. However, this stage is not usually encountered in practice as the product will be well and truly rancid by this time and probably altered in physical characteristics.

An antioxidant works by reacting with the free radicals as they are formed, converting them back to the original substrate. Free radicals of the antioxidant molecules are formed in this process but the structure of antioxidant is such that these are relatively stable and do not have enough energy to react with the fat to form further free radicals. It follows that an antioxidant will only really be effective if it is added during the initiation period. Figure 5.2 illustrates the effect of adding an antioxidant at zero time to a fat. It is an idealised fat, but the relative induction periods are typical for lard without an antioxidant and with the antioxidant BHA added.

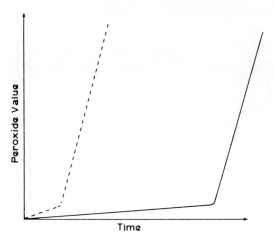

Figure 5.2 Effect of adding an antioxidant to lard. ---, unstabilised fat; ———, stabilised fat.

If the antioxidant is added later during the induction period, the extension in shelf life will be much less noticeable. Figure 5.3 illustrates this in a graphical manner, again using an idealised fat. This leads to one of the first principles of the use of antioxidants, no matter which antioxidant is being used. The antioxidant should be added to the fat as early as is possible in its life to produce the maximum effect. With a refined deodorised vegetable oil this stage is immediately after the deodorising processing has been completed and the oil has cooled to about 60–80 °C.

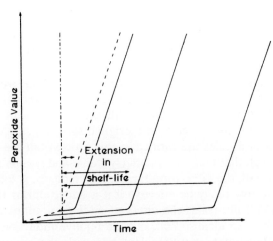

Figure 5.3 Effect of adding an antioxidant at various stages during induction period. ---, unstabilised fat; ———, stabilised fat.

5.5 Synergism

In some circumstances it is found that a combination of two or more antioxidants works better than the equivalent quantity of any one antioxidant. This is known as synergism.

Figure 5.4 illustrates the effect of using BHA and BHT in combination in lard. The reason why combinations of antioxidants work more efficiently than a single antioxidant is not really known, but there are a number of free radical reactions occurring during the oxidation process and it is possible that antioxidants differ in their efficiency in dealing with the different mechanisms. Not all combinations of antioxidants display synergism. More details are given in section 5.7 on individual antioxidants.

A number of acidic compounds, including citric acid, phosphoric acid, ascorbic acid and EDTA, help to increase the shelf life of a fat when used in conjunction with an antioxidant. These work by complexing with trace quantities of metals which would otherwise promote the oxidative reaction. These compounds are often known as sequestering agents. They are not true antioxidants and do not exhibit any antioxidant activity in a fat from which all traces of metal have been removed. These sequestering or complexing agents are also sometimes misnamed as 'synergists'. Citric acid is the most commonly used sequestrant, either in antioxidant formulations and/or, after deodorising, in vegetable oil refining.

The value of sequestrants should not be overlooked. It has been found that only very small traces of certain metals have to be present for a marked deterioration in the keeping quality of a fat. Iron will have an adverse effect from concentrations as low as 0.01 ppm.

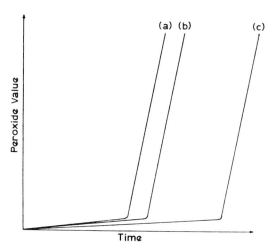

Figure 5.4 Synergism in typical animal fat, (a) with 0.02% BHA, (b) with 0.02% BHT, (c) with 0.01% BHA + 0.01% BHT.

5.6 Popular misconceptions

There are a number of misconceptions about what an antioxidant can and cannot do. Most of these should follow from an understanding of the mode of action of an antioxidant, but it is worth emphasising some points. An antioxidant cannot:

1. Improve the flavour of poor quality fats or oils.
2. Improve an oil in which oxidative rancidity has already developed.
3. Prevent microbial decay.
4. Prevent hydrolytic rancidity.
5. Prevent ketonic rancidity.
6. Prevent flavour reversion.

At the very best an antioxidant can only be expected to maintain an oil in its original condition. It cannot be expected to improve the flavour of a poor quality fat or oil or to improve an oil in which oxidative rancidity is already pronounced.

An antioxidant cannot be expected to prevent microbial decay. However, having said this, many antioxidants are phenolic in nature and can be shown to have some antimicrobial activity. BHA in particular has quite marked antimicrobial activity,[2] probably of a similar order to the parabens (*p*-hydroxybenzoate) esters, against *Aspergillus parasiticus* (aflatoxin producer), *Escherichia coli*, *Salmonella typhimurium* and *Staphylococcus aureus*. The greatest activity is against moulds and Gram-positive bacteria.

Studies[3] on BHT showed it to be slightly inhibitory at 1% to *Salmonella senftenberg*. However, it is difficult to employ the antimicrobial activity of BHA or BHT in most applications because they are virtually insoluble in water.

Antioxidants will not prevent the formation of free fatty acids, which are produced by the chemical hydrolysis of fat and give rise to a form of rancidity known as hydrolytic rancidity. The problem is often found in high temperature frying processes where water is present and there is no additive which will effectively prevent this. Hydrolysis of fat and the production of free fatty acids may also be encountered with crude fat, where enzymes which promote this hydrolysis may be present. The coenzymes are released from the cell wall of the seed or plant by crushing or bruising during the extraction process.

5.7 Properties of the common antioxidants

This section deals with the characteristics of the various antioxidants used in fats and oils that are permitted by European regulations or other authorities. The form of the regulations affects the way in which they are used.

5.7.1 Tocopherol

Figure 5.5 shows the characteristics of tocopherol and ascorbyl palmitate. Although they are covered in chapter 6, we need to consider the tocopherols in the context of the use of synthetic antioxidants, as most of the product sold for this use is synthetic. The structure in Figure 5.5 is that for α-tocopherol (vitamin E), but γ and δ tocopherols are also found. The tocopherols are widely distributed throughout the vegetable kingdom and have been called 'nature's antioxidants'. They are present in appreciable quantities in all vegetable oils, levels up to 3000 ppm being encountered. Fats derived from animals and fish contain virtually no tocopherol and this is considered to be the reason for their much poorer stability compared with vegetable oils.

An appreciable quantity of the natural tocopherol content of an oil survives the refining and deodorising process, although this can vary with process conditions. There is little benefit to be gained from adding additional tocopherol to the majority of vegetable oils to improve their storage life. It is a case of nature providing the optimum level. However, the natural tocopherol content seems less capable of surviving a number of common food processing operations. Certainly there is a marked change in the stability of a vegetable oil before and after its use in a deep frying operation and supplementing the natural antioxidant with a synthetic may well be beneficial.

5.7.2 Ascorbyl palmitate

Ascorbyl palmitate is often referred to as a natural antioxidant but strictly speaking this is not correct. Ascorbic acid (vitamin C) occurs widely in the vegetable world; the palmitoyl ester does not occur naturally. Ascorbic acid and its salts are virtually insoluble in fats and oils and consequently cannot in practice be used as antioxidants in lipids. Being water soluble they are used as antioxidants in products such as beer, soft drinks and fruit juices. Ascorbyl palmitate is synthesised because it has a greater solubility in oil. However, its solubility is still very low compared with other antioxidants and in order to dissolve ascorbyl palmitate it is usually necessary to use it in combination with a solubilising agent, such as a monoglyceride. Even so, it may be necessary to resort to high temperatures to achieve an acceptable rate of dissolution. It is still a matter of debate as to whether ascorbyl palmitate is a true antioxidant. One school of thought believes that it is not a true antioxidant and only shows antioxidant activity by virtue of its ability to act as a sequestering

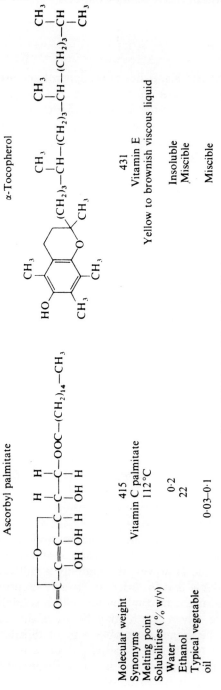

Figure 5.5 'Natural' food antioxidants.

	BHA	BHT	Gallates	TBHQ
	OH, C(CH₃)₃, OCH₃	OH, (CH₃)₃C, C(CH₃)₃, CH₃	OH, HO, OH, COOR; R = C₃H₇, propyl; R = C₁₂H₂₅, dodecyl	OH, C(CH₃)₃, OH
Molecular weight	180	220	212—propyl; 338—dodecyl	166
Synonyms	Butylated hydroxyanisole 3-*tert*-Butyl-4-hydroxyanisole 2-*tert*-Butyl-4-methoxyphenol	Butylated hydroxytoluene 3,5-Di-*tert*-Butyl-4-hydroxytoluene 2,6-Di-*tert*-Butyl-*p*-cresol; DBPC 2,6-Di-*tert*-Butyl-4-methylphenol	Dodecyl = Lauryl Propyl/Dodecyl-3,4,5-trihydroxybenzoate	Mono-*tert*-butylhydroquinone
Melting point	50–52°C (commercial food grade)	69–70°C	Propyl, 146–148°C; Dodecyl, 95–98°C	126.5–128.5°C
'Carry-through' properties	Very good	Fair–Good	Propyl, Poor; Dodecyl, Fair–Good	Good
Synergism	Yes with BHT and gallates	Yes with BHA but not gallates	Yes with BHA	
Solubilities (% w/v)			Propyl / Dodecyl	
Water	0	0	0.35 / 0.0001	1
Propylene glycol	50	0	65 / 4	30
Typical animal fat	30–40	20–30	1 / 1	5–10
Typical vegetable oil	40	20–30	1 / 1	5

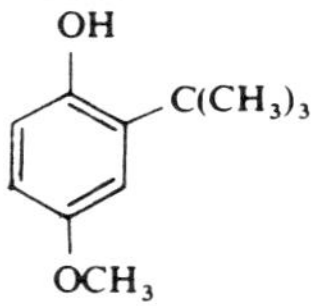

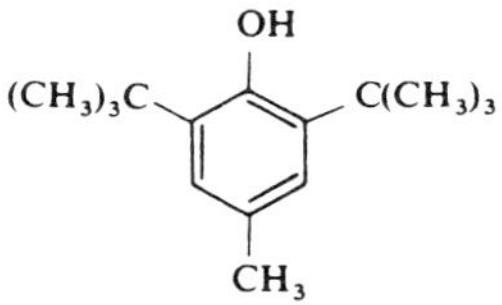

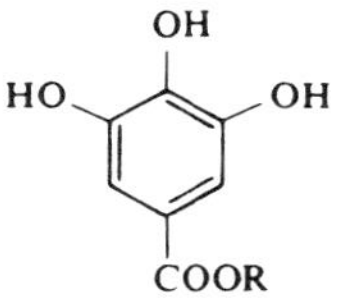

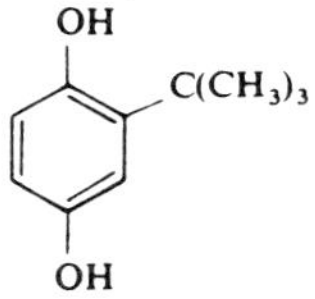

Figure 5.6 Synthetic food antioxidants.

agent. It is seldom used on its own, normally being used in a formulation with the tocopherols.

5.7.3 BHA

Butylated hydroxyanisole (BHA) is shown in Figure 5.6. BHA has a high solubility in animal fats and vegetable oils and is a very effective antioxidant for animal fats. BHA has good stability under the conditions encountered in many food processing operations and gives good stability to a fried or baked product. This 'carry-through' property of BHA is a very important consideration in the choice of an antioxidant. The effect of BHA is much less marked in a vegetable oil which has an adequate natural tocopherol content. This applies to the oil being stored without any further processing. If the oil is to be used for a food manufacturing process, such as deep fat frying, then the carry-through comes into play and BHA will show a considerable effect on the stability of the final product. This is illustrated in Figure 5.7. These results are taken from unpublished work conducted at the Leatherhead Food Research Association, an independent laboratory. In a typical commercial frying oil, an equal mixture of hardened soyabean oil and palm olein, BHA had

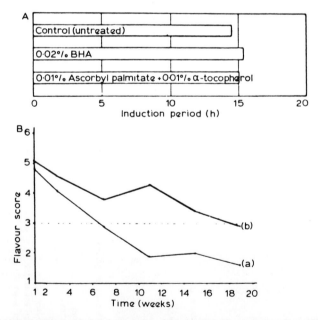

Figure 5.7 (A) Induction periods of 1:1 palm olein: hardened soyabean oil at 100 °C, using FIRA–Astell apparatus. (B) Flavour score of potato crisps stored at 30 °C, after frying under simulated commercial conditions in 1:1 palm olein:hardened soyabean oil: (a) untreated control, (b) oil containing 0.02% BHA initially. A flavour score of 6 is bland with no trace of off-flavours, 3 is borderline for acceptability, 1 is rancid.

virtually no effect on the induction period of the oil. However, it greatly improved the keeping quality of the crisps which were fried in this oil under simulated commercial conditions. The simulation subjected the oil to the quantities of water and high temperature which it would experience in a commercial crisp frier, by adding water slowly into the oil for 24 h while maintaining it at 180 °C. At the end of this period a batch of crisps was fried and kept at 30 °C. One disadvantage of BHA is its steam volatility at frying temperatures. It is often claimed that so much is lost by steam distillation that it can have little effect on the final product. This is not so. At the end of this particular test there were only 11 ppm of BHA left of the initial 200 ppm but even this low level is clearly effective. By protecting the oil during frying the BHA has done its job and the result is a longer shelf life, irrespective of the fate of the antioxidant. An ascorbyl palmitate and α-tocopherol mixture in the same oil blend was found to be no more effective than BHA in extending the induction period, i.e. the initial oil storage stability (Figure 5.7). Another important property of BHA is that it exerts a synergistic action with both BHT and the gallate esters.

5.7.4 BHT

In many ways BHT (butylated hydroxytoluene) is very similar to BHA. It is very soluble in fats and oils, although not quite as soluble as BHA. Unlike BHA it is insoluble in propylene glycol, which poses some problems in introducing a formulation containing BHT and citric acid. Citric acid is often used in formulations as a sequestering agent, but it has a low oil solubility. BHT is very effective in animal fats, but much less effective in vegetables oils. It is able to survive the high temperatures encountered during food processing operations, but is not as good in this respect as BHA. It does not show any synergistic action with the gallates but, as already mentioned, is synergistic with BHA.

BHT is more steam volatile than BHA. As BHT is often used in combination with BHA this can pose particular problems in some food processes. In the manufacture of dehydrated potato products, for example, a BHA/BHT combination is very effective in retarding autoxidation of the flavour lipids, but use of a traditional one-to-one mixture of BHA/BHT results in almost total loss of BHT due to steam distillation. This loss is often allowed for by the addition of about 5 times as much BHT as BHA in the additive mix.

5.7.5 Gallate esters

The propyl, octyl and dodecyl gallates (see Figure 5.6) are permitted in Europe but in the UK only propyl gallate is used in any quantity. Propyl gallate is an effective antioxidant for both animal fats and vegetable oils. All the gallates are much less soluble in fat or oil than BHA or BHT. They also suffer from the disadvantage that they are to varying degrees sensitive to heat and do not survive as well in most cooking processes. Since the temperature stability increases with molecular weight, propyl gallate is worst in this respect and dodecyl gallate is the most stable. Gallates are often employed in combination with BHA, the gallate giving good initial stability to a vegetable oil and the BHA ensuring that the cooked product has a good shelf life. This practice is recognised in regulations – gallates are used in liquid formulations to overcome the problems of low solubility.

All the gallates have the disadvantage of forming blue-black reaction products ('inks') with iron in the presence of water. The lower molecular weight gallates are worse in this respect than those with a higher molecular weight. The tendency to produce discoloured products can be reduced by the use of a suitable sequestering agent, e.g. citric acid.

This colour development can reach serious proportions in oil storage, the application where propyl gallate shows to best effect. Frequently bulk oil storage tanks collect water 'bottoms' and, especially if they are of mild steel construction, corrosion means the presence of dissolved or suspended iron. The propyl gallate is preferentially soluble in the water bottom and reacts to form a black discoloration which then affects the colour of the oil. Mishaps of this type have caused users to switch away from an otherwise very effective antioxidant for oil storage and draw attention to the need to maintain plant/tanks correctly.

5.7.6 TBHQ

The final antioxidant outlined in Figure 5.6 is TBHQ (t-butylhydroquinone). This is the newest food antioxidant in common use, but even this has a long history having gained its first approval for food use in 1972 in the USA. At present it is not permitted in the UK or the EC but has been approved by authorities in many other countries. This antioxidant fills the gap in the existing range. It is a very effective antioxidant for vegetable oils, including the unsaturated vegetable oils, and is even more effective than the gallates. Unlike the gallates it does not suffer from a poor oil solubility or the tendency to discolour in the presence of iron and water. In addition, it is stable at high temperatures and is slightly less volatile than BHA and BHT. This would lead one to expect superior carry-through properties but these expectations are, unfortunately, not met. TBHQ has fairly good carry-through properties, showing a similar effect to BHA in fried foods. For baked foods it appears to be less effective than BHA; this may be a pH effect.

Figure 5.8, again based on our work carried out at the Leatherhead

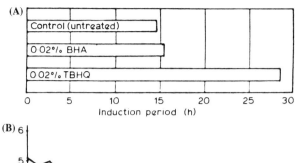

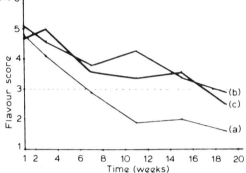

Figure 5.8 (A) Induction periods of 1:1 palm olein: hardened soyabean oil at 100 °C using Fira–Astell apparatus. (B) Flavour score of potato crisps stored at 30 °C, after frying under simulated commercial conditions in 1:1 palm olein:hardened soyabean oil: (a) untreated control, (b) oil containing 0.02% BHA initially, (c) oil containing 0.02% TBHQ intially. For definition of flavour scores see caption to Figure 5.7.

Food Research Association, illustrates the effectiveness of TBHQ in extending the induction period of fresh oil and in extending the shelf life of crisps prepared under simulated commercial conditions. (The results for BHA here are the same as in Figure 5.7.)

TBHQ has been shown to be particularly effective in stabilising crude oils during storage and improves the quality of the refined and deodorising oils obtained from them, even though the refining and deodorising process removes all the TBHQ. This is of value to countries producing and exporting crude vegetable oils, such as palm oil from Malaysia and Indonesia, since because of the shipping distances involved the product is stored for a considerable period of time.

5.8 Carry-through

Several references have been made to the ability of various antioxidants to survive cooking processes, referred to as their 'carry-through' property. Figure 5.9 illustrates the effect obtained with BHA, BHT and propyl gallate in a lard used to make pastry. BHA is the most effective single

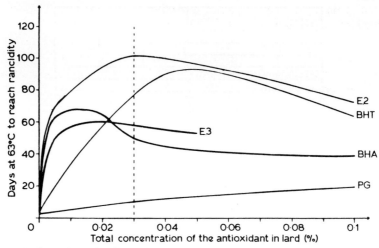

Figure 5.9 'Carry-through' effects of antioxidants in pastry made with lard (Schaal oven, 0.01% antioxidant).

antioxidant for concentrations in the range 100–200 ppm, which is the normal maximum permitted by most regulations. Unlike BHA, BHT shows increasing effect as the concentration increases. The poor effect (poor carry-through) of propyl gallate is clear. A mixture of BHA and BHT ('Embanox' 2*) is better than either antioxidant used alone. E.3 ('Embanox' 3) is a BHA/propyl gallate/citric acid formulation. The

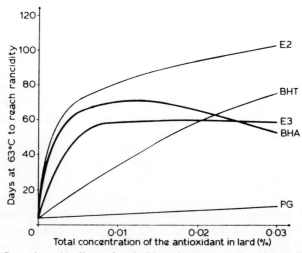

Figure 5.10 'Carry-through' effects of antioxidants in pastry made with lard (Schaal oven, 0–0.03% antioxidant).

regulations in most countries allow a maximum of 100–300 ppm of these antioxidants, and Figure 5.10 illustrates this region in more detail.

This is the area in which selection of an antioxidant is focused, as the efficacy at levels above the permitted maximum is of no more than academic interest. The very steep slope for BHA illustrates its surprising efficacy at very low residual concentrations.

5.9 Practical aspects of using antioxidants

Producing a uniform mixture of substances used in similar proportions is often not too difficult but when one component is required in only a very low concentration, as low as 100 ppm, special care is required. This is the first of the practical aspects and difficulties encountered when using antioxidants. We have already noted that to be used to best effect the antioxidant must be dispersed evenly throughout the foodstuff, usually a fat or oil. The most common method of using an antioxidant is to add it directly to a fat or oil. The concentration of antioxidant is very small and it may prove difficult to ensure that it is evenly dispersed. Clearly this can only be done with the fat/oil in a liquid state, and maintaining the oil at 60–80 °C will usually give sufficient mobility to ensure uniform dispersion of the antioxidant. With vegetable oils this means immediately after deodorisation. The problem is eased somewhat if a liquid formulation is used, such as 'Embanox' 5, a 20% solution of BHA in vegetable oil, or if the antioxidant is dissolved in a small quantity of the oil to be stabilised. Even so, it will be necessary to have efficient agitation to ensure that it is evenly dispersed, but this agitation should not be so violent as to draw air into the oil by vortexing. The difficulties encountered in obtaining an even dispersion of the antioxidant is one of the reasons for recommending a potential food manufacturing user to purchase stabilised oil from the refiner who is the best equipped to incorporate an antioxidant. Some suppliers of vegetable oil processing plants incorporate an antioxidant addition system in plant design. This consists of a small tank and a metering pump which will inject the deodoriser to the storage tank. This is ideal; the oil is hot, mobile and turbulent, and gets thoroughly mixed as it fills the storage tank. Antioxidant addition prior to delivery, e.g. on transfer from storage to road tanker vehicle, is second best; the mixing is less thorough.

Another common method of using antioxidants is spraying a solution of the antioxidant on to the foodstuff as it passes by on a conveyor belt before packing. Uniform dispersion can be a problem but this is not too bad if the foodstuff is being packed into sealed packs which are stored for a period of time. BHA and BHT have an appreciable volatility, even at room temperature, and tend to disperse evenly over a fairly short period in the fat or oil.

5.10 Typical applications of antioxidants

Although continued reference has been made to the use of antioxidants in 'fats and oils', these are frequently only an ingredient of the final food product. Table 5.1 is included to illustrate the very wide range of products in which antioxidants are found to be useful. New regulations (a 'Directive') are awaited in the European community which would specify permitted applications. The wording of the US and other national regulations permits a wider range of use. In the UK the regulations have been interpreted as only permitting addition 'in' certain fats, oils, etc., which precludes 'topping-up' antioxidant lost in processing, say, snack foods, with antioxidant in the salt/flavour in the surface coating applied after cooking. Such use is common in the US and elsewhere and seems sensible from the technological justification viewpoint.

Table 5.1 Typical applications of antioxidants

Animal fats—lard, dripping and suet (block and shredded)[a]
Vegetable oils[a]
Hydrogenated vegetable oils[a]
Shortenings[a]
Butter (for manufacturing)[a]
Margarine[a]
Fish and fish-liver oils[a]
Biscuits and shortbread[a]
Cake mixes (non-fat)
Baked products[a]
Potato crisps[a]
Reformed starch snack products (fried)[a]
Fried nuts[a]
Cereal products
Chewing gum base[a]
Dried soups[a]
Dairy products (for manufacturing)[a]
Dehydrated potato products[a]
Dried vegetables
Vitamin preparations—A and D[a]
Essential oils[a]
Citrus oils[a]
Emulsifiers and stabilisers (containing combined fatty acids)[a]
Cosmetics
Waxes and tallows
Meat and bone meal
Fish meal
Food packaging

[a]Indicates applications permitted by UK regulations.

5.11 Permitted rates of use

Table 5.2 gives the permitted rates of use in fats and oils in the UK, the EC and the USA. These do change, albeit rarely, and the list is not intended as the basis for answering regulatory queries. The table has been arranged in similar groupings. It is interesting to note that both Belgium and The Netherlands allowed for the fact that some antioxidant is lost during manufacturing processes by allowing a double limit in fats/oils intended for use in manufacturing. Germany has been the exception but will come in line with the EEC with adoption of the new EC Directive.

Generally, food labelling law in the developed countries requires the declaration of most food additives and in the case of antioxidants this may take the form 'permitted antioxidants' or specific or general name with/without 'E. number'. Local regulations should be checked.

5.12 The future

What of the future of antioxidants? Despite the 'all additives are harmful' lobby and their insistence that we revert to 'natural' foods, the fact is that more and more people are eating more and more processed foods. Their manufacture is possible only by using additives, and this implies a slow but steady growth in the demand for antioxidants. TBHQ has still not received EC approval, which would open the door to a much wider use of this antioxidant, leading to an increase in the use of antioxidant mixtures

Table 5.2 Permitted rates of use of antioxidants in fats and oils in the UK, the EC and the USA. New regulations pending in EEC, possible new levels are indicated in brackets

	Antioxidant			
	BHA	BHT	Gallates	TBHQ
USA	200	200	150	200
UK	200	200	100	–
Eire	200	200	100	–
Belgium—retail	100	100	100	–
manufacturing	400	400	400	–
Netherlands—retail	100	100	100	–
manufacturing	400	400	400	–
Italy	300	300	100	–
France	100	100	100	–
Luxembourg	100	100	100	–
Denmark	100	100	50	–
West Germany				
[EEC – likely new levels:	200	100	200	]

or formulations. This analysis assumes that the anti-additive hysteria seen in the UK eventually will moderate as consumers realise the risks associated with the 'natural' degradation of foodstuffs on storage and alarming effects as those attributed to manufactured additives.

After TBHQ, what next in synthetic antioxidants? One candidate from the USA was the polymeric antioxidant. Designed to allay all fears of unforeseen toxic reactions, this product was designed to pass through the human gut without being absorbed. It is a high molecular weight polymer with phenolic side-chains having the structures of BHA/BHT/TBHQ/HQ derivatives. As the active part of the molecule is relatively small, it is necessary to use a lot more antioxidant to achieve an acceptable level of activity.

Figures 5.11 and 5.12, taken from work commissioned by May & Baker Ltd. at the Food Research Association, illustrate the activity of the polymeric antioxidant. In crisp frying oil, only a slight effect was obtained at 200 ppm. In sunflower oil (a polyunsaturated vegetable oil) a more

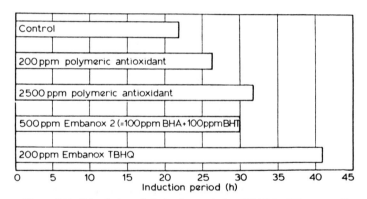

Figure 5.11 Crisp frying oil, induction period at 100 °C by FIRA–Astell.

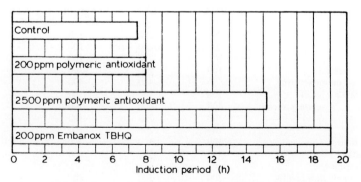

Figure 5.12 Sunflower oil, induction period at 100 °C by FIRA–Astell.

significant effect was shown at 2500 ppm, but was still inferior to 200 ppm TBHQ. At the time of writing this product has been submitted for clearance in the USA but has not received FDA approval.

Another area receiving attention is the antioxidant properties of as yet not completely identified components of some spice extracts, rosemary in particular (see chapter 6). Initial evidence suggests that these extracts are less effective than the better antioxidants currently used, but it is possible that identification of the active principles, such as the triterpene carnosol, may lead to synthetic manufacturing routes enabling production of a cost-effective antioxidant. It will of course be necessary to demonstrate the safety of such products, as it cannot be assumed that they present no hazards simply because they occur in nature.

Beyond this it is difficult to see any major new antioxidant products on the horizon. The cost of conducting the toxicity studies required of a new food additive are prohibitive compared with the size of the potential market. Some large companies have lost money on attempted innovations in this field, and the half-dozen products currently in use will probably have to suffice for some time to come.

Acknowledgement

The author wishes to acknowledge the assistance of P. A. White, BSc in the preparation of this manuscript and in the design of the May & Baker Ltd. experimental work described.

References

1. Tollenaar, F. D. and Vos, H. J. (1958). *J. Am. Oil Chem. Soc.*, **35**, 448–55.
2. Chang, H. C. and Branen, A. L. (1975). *J. Food Science*, **40**, 349.
3. Ward, M. S. and Ward, B. Q. (1967). *Poultry Sci.*, **46**, 1601.
4. Williams, G. M., Sies, H. and Baker, G. T. (1993). *ANTIOXIDANTS—Chemical, Physiological, Nutritional and Toxicological Aspects*, American Health Foundation, Valhalla, New York.

6 Spectrophotometric and chromatographic assays
E. PRIOR and J. LÖLIGER

6.1 Introduction

Lipid oxidation is a complicated set of autocatalytic processes leading to unacceptability of a food, usually by the development of unpleasant tastes and flavours. But it also includes loss of vitamins and other essential nutrients (e.g. polyunsaturated fatty acids) and loss of charcteristic aroma. The end-result is decreased quality and reduced shelf life of the product. In some situations health-related issues of food are associated with an advanced state of oxidation of polyunsaturated lipids.[1] For a detailed discussion of the initial steps of lipid oxidation refer to chapter 1.

As oxidation progresses the number of products increases dramatically, e.g. oxidation of linoleic acid (one substrate) yields four hydroperoxides and their decomposition results in 100 or more secondary products (Table 6.1). At present no one universal technique exists for determining the state of oxidation and the food scientist needs to use a range of methods.

A wide range of techniques can be used for determining lipid oxidation, but this contribution will concentrate on chromatographic and spectrometric methods and in particular their application to foods. Rather than cataloguing the methods we have chosen a thematic approach based on the pathway of oxidation. The methods are summarised in Table 6.2.

Table 6.1 Secondary oxidation products from linoleic acid[2]

Product	Amount $(\mu g\,g^{-1})^a$
Pentane	$+^b$
Hexanal	5100
2-t-Heptenal	450
2-c-Octenal	990
2-t-Octenal	420
2-t-4-c-Decadienal	250
2-t-4-t-Decadienal	150

^{a}Compounds greater than 100 $\mu g\,g^{-1}$ are reported. ^{b}Pentane was not quantitated under the experimental conditions, but is an important and easily measured oxidation product.

Table 6.2 Summary of methods available for following oxidation

Primary reactions/products	determined by:
Substrate	Fatty acid composition
	Oxygen consumption
Radicals	ESR
Hydroperoxides	Peroxide value
	Conjugated dienes
	HPLC
	GC as TMS-esters
	Indirect methods e.g. pyrolysis
	Chemiluminescence
Secondary products	
Volatiles	TBARS
	p-Anisidine value
	Carbonyls e.g. as 2,4-DNPH by spectophotometry, HPLC or GC
	Direct headspace sampling
	Purge and trap
	Simultaneous distillation extraction
Fluorescent products	Fluorescence techniques

6.2 Substrate

Determination of the fatty acid composition as their fatty acid methyl esters (FAMEs) by gas chromatography (GC) is a commonly used technique in the study of lipids. Since fatty acids are the basic substrate of lipid oxidation it would seem logical that their consumption should correspond to the progress of oxidation. However, this technique is not very sensitive: lipids can become rancid at very low oxidation levels ($< 0.00002\%$) and it is analytically extremely difficult to measure such small differences in large amounts. For fatty acids present in large quantities, e.g. more than 20%, the precision for FAMEs analysis is ca. 0.1–0.5%, and for fatty acids present in much smaller quantities the precision falls to 5–10%.

The other substrate for oxidation is oxygen. For obvious reasons its consumption must be measured in a sealed container. Oxygen in a headspace sample is commonly measured, along with other gases, on a molecular sieve column using a thermal conductivity detector. However, analysis time can be long (up to 1 h) due to interference from water. A more rapid and elegant solution is to measure oxygen content with a paramagnetic detector, which can be performed at the same time as sampling for headspace gases (see section 6.5.3.1). The same constraint applies to oxygen determination as to fatty acids: it is difficult to measure small differences in large numbers. Nevertheless, as the measurement of residual oxygen in the headspace requires very little sample preparation it is easily applied to complex food systems where extraction (clean-up and analysis) of the sample is difficult.

6.3 Radicals

Radicals involved in lipid oxidation are most commonly peroxyl, alkoxyl, hydroxyl or alkyl. These short-lived species are normally measured by electron spin resonance (ESR) at low temperatures or as spin adducts. The interpretation of their structure is used for elucidation of the mechanism of oxidation and antioxidation reactions.

Free radicals are ubiquitous in foods and can easily be observed in dry foods at room temperature. Transition metal ions are also commonly found in foods. For instance, recent studies have demonstrated the presence of organic free radicals, Mn^{2+} and Fe^{3+} in cereals.[3] Figure 6.1(a) shows the progression of the ESR signal measured semi-quantitatively during storage of oat flours either non-treated, heated under pressure or roasted. Figure 6.1(b) shows the development of headspace pentane in these samples (P. Lambelet, 1990, private communication). At present all the radicals in foods cannot be identified, but this rapid and non-destructive technique offers much promise for the future.

6.4 Hydroperoxides

6.4.1 Spectrophotometric methods

Hydroperoxides are known as the primary products of lipid oxidation and, although transient, are much more stable than the radicals. Linoleic and linolenic acid hydroperoxides formed by autoxidation contain a conjugated diene group which has a strong absorbance at 234 nm; this characteristic forms the basis of their spectrophotometric determination. Their large molar absorptivity (26 000) means that dilute solutions are adequate. The conjugated diene value can be measured by dissolving the lipid in cyclohexane or iso-octane (0.05%) and measuring against methyl stearate (1%). This reference is needed bcause of the triglycerides' strong absorbance at ca. 190–210 nm, due to functional groups such as carboxyl, esters and ethylene, which form an interfering shoulder in the 234 nm region and decrease sensitivity. This method is principally used for determining the degree of refining (bleaching) of an oil as this process produces dienes, trienes and tetraenes (absorbance at 234, 274 and 392 nm, respectively) not normally present. However, this type of analysis has the advantage of being non-destructive and the authors use it (without a methyl stearate reference) to monitor the production of

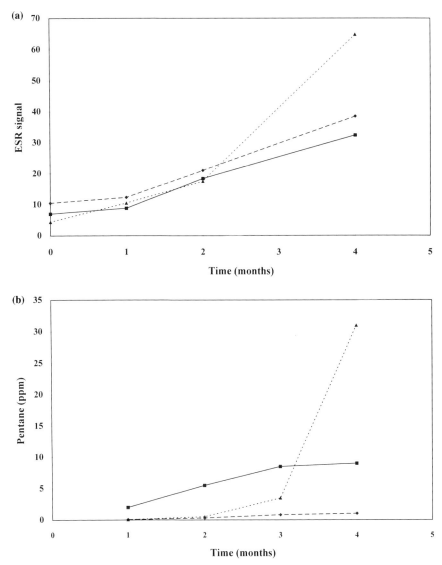

Figure 6.1 Results of shelf-life test at 37 °C of oat-based cereal: non-treated (■), heated under pressure (◆) and roasted (▼). (a) ESR signal and (b) headspace pentane.

hydroperoxides from linoleic acid by lipoxygenase. It should be noted that quantification of conjugated dienes cannot be used as an absolute measure of lipid oxidation: dienes are found in lipid alcohols derived from peroxides and in certain non-oxidised fatty acids. Also, not all hydroperoxides possess a conjugated diene group, e.g. oleic acid hydroperoxides. Although measurement of conjugated dienes is simple and gives a rapid indication of the extent of oxidation for determination of hydroperoxides we recommend using more specific and/or more sensitive techniques.

In the context of spectrophotometric determination of oxidation products, the COP assay (conjugable oxidation products)[4] should be mentioned, in which hydroperoxides from polyenoic acids are reduced and then dehydrated to more conjugated chromophores. It can distinguish between oxidation of linoleic acid and of the more highly unsaturated fatty acids. The first step, reduction by sodium borohydride, results in the disappearance of the characteristic UV absorbance at 275 nm of the oxodienes which is defined as the oxodiene value. Although this method has been widely reported in reviews of lipid oxidation methodology it has not been widely applied.

The peroxide value (PV) is classically determined by titration of the iodine released from the reaction of peroxides with potassium iodide. This can be easily automated with electrochemical determination of the end-point which eliminates problems due to colorimetric determination (e.g. poor sensitivity at low PVs, and coloured samples). The principal disadvantage is the amount of sample required, 1–5 g. For oil samples relatively large quantities are normally available, but for lipid samples extracted from food the amount available can be very limited, often less than 100 mg. A technique routinely used in our laboratories in this situation is spectrophotometric determination at 510 nm of the red thiocyanate complex formed by reaction of excess sodium thiocyanate with ferric iron oxidised from ferrous iron by the peroxides.[5] This method is particularly well adapted to measurement of PV in foods, as the dichloromethane–methanol lipidic solution obtained after cold extraction can be used directly without further treatment. Soxhlet extraction of lipids (at high temperatures for several hours) gives a large error in the determination of PV: high temperatures lead to decomposition of hydroperoxides on the one hand and heating for a long period forms additional hydroperoxides on the other. This colorimetric method does have certain limitations due to interference from pigments in the lipid extract, e.g. from cured meats and dehydrated tomato products, which give a large signal at ca. 510 nm.

6.4.2 HPLC methods

To overcome the problem of specificity high performance liquid chromatography (HPLC) techniques can be used. HPLC has been extensively used for the separation and determination of hydroperoxides (and other peroxides, e.g. diperoxides, epidioxides) for the elucidation of the mechanism of oxidation and, for example, in studies on lipoxygnase. It has the great advantage of operating at room temperature or lower, which decreases the risk of hydroperoxide decomposition. It has been frequently used to determine oxidation in biological samples; however, there is relatively little literature available on its application to the determination of oxidation in fats and oils[6] and in particular in foods.

The authors have developed several HPLC techniques, the simplest of which is to dissolve the sample (fat, oil or lipid extract) in hexane (2%) and to separate the oxidised triglycerides from the non-oxidised triglycerides by normal phase chromatography. Under our routine conditions, 0.9% isopropanol in hexane at 1 ml min^{-1} with a Nucleosil 100-3 column (125 mm × 8 mm × 4 mm (Macherey-Nagel, Oensingen)) and a UV–vis detector at 234 nm, the analysis is complete in 10 min. Addition of a fluorescence detector allows the simultaneous determination of tocopherols. This analysis is particularly useful for small samples, since using an automatic injector, as little as 200 μl of solution are required (4 mg lipid). Also samples containing interfering substances, e.g. ascorbic acid, falsify the electrochemical determination of the end-point in iodometric analysis and pigments in the ferric isothiocyanate analysis.

Figure 6.2 shows the development of triglyceride hydroperoxides (TG-OOH) as measured by HPLC during the oxidation of blackcurrant seed oil containing a variety of antioxidants. It should be noted that use of a UV detector at 234 nm implies that only the conjugated dienes formed by autoxidation of linoleic and linolenic acid will be measured and since

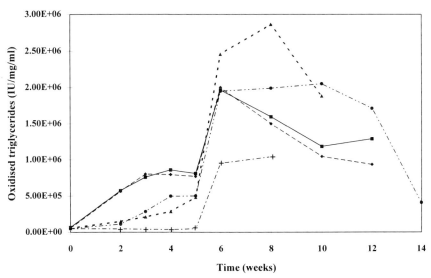

Figure 6.2 Progression of oxidation in blackcurrant seed oil stored at 50 °C followed by HPLC of TG-OOH. Oil treated with various antioxidants: non-stabilised (■), ascorbyl palmitate, 200 ppm (♦), Nipanox, 4000 ppm (▲), ternary antioxidant, 9000 ppm (+) and quaternary antioxidant, 20 000 (●).

these are the most important hydroperoxides formed in foods, their analysis is in general sufficient. Correlation of HPLC with PV for oils from a number of canola varieties is shown in Figure 6.3; the correlation is linear up to a certain level, above which the increase in PV is more than the increase in TG-OOH due to formation of other peroxide-containing compounds. Application to soy-based powders from different sources with and without addition of trace minerals is shown in Figure 6.4. This HPLC method is simple, rapid and lends itself to routine analysis with automatic injection. However, under the above conditions the individual hydroperoxides cannot be separated (although separation of TG-OOHs can be achieved with a longer run-time). Also more polar lipids, i.e. phospholipids, are trapped on the column and are not measured.

To determine individual hydroperoxides a second HPLC technique has been developed based on analysis of the hydroperoxides (from free fatty acids, triglycerides or phospholipids) as their hydroxy fatty acids. The lipid sample (50 mg) is dissolved in ethanol (0.5 ml) and iso-octane (0.5 ml) and 1 ml of saturated sodium borohydride in isopropanol is added. The solution is heated at 60 °C for 60 min in a heated block and the solvent removed under vacuum. The residue is taken up in hexane (3 ml) and sonicated to ensure full dissolution. The precipitate is brought down by centrifugation (1 500 rpm/5 min) and the supernatant analysed by normal phase chromatography under the same conditions as for the TG-OOHs above. Reduction of the hydroperoxides to hydroxy derivatives increases their stability, and the pH of the reaction mixture hydro-

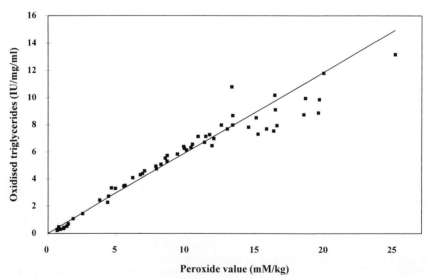

Figure 6.3 Correlation between TG-OOH measured by HPLC and peroxide value for canola oils. $y = 0.595x + 0$, $r = 0.969$, $n = 50$.

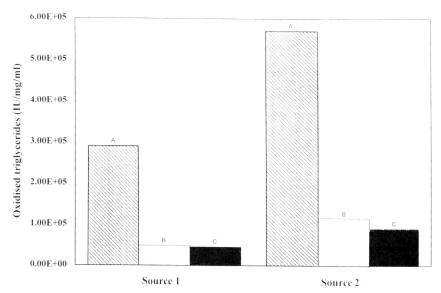

Figure 6.4 TG-OOH measured by HPLC in soya-based products from two different sources: (A) soya powder, (B) soya powder with maltodextrin and (C) soya powder with maltodextrin and trace elements.

lyses the triglycerides and phospholipids to their fatty acids. The geometric isomers of the resulting hydroxy fatty acids (as ethyl esters due to presence of ethanol) can be well separated by HPLC. Figure 6.5 shows the development of hydroxy fatty acids during oxidation of a number of oils. Values reach a maximum and afterwards decrease, which is typical of hydroperoxides as they are unstable intermediates. The increase corresponds to production > destruction, the plateau to production = destruction and the decrease to production < destruction.

The correlation of HPLC values to PV is shown in Figure 6.6. As with TG-OOHs, the relationship is linear up to a certain level and afterwards falls off. It should be emphasised that the aim of these two HPLC techniques is not to replace the PV; the values obtained are relative and depend on the fatty acid composition of the oxidising lipid. Rather, they provide a complementary method, especially for small samples, samples which cannot be analysed by classic methods and for obtaining additional information, e.g. proportions of geometric isomers. Application to the shelf life of salami containing different antioxidants during frozen storage is shown in Figure 6.7, where other methods could not be used due to small sample sizes and the presence of pigments in the lipid extract.

The sensitivity of the method can be increased ca. 10–20-fold by purification of the hydroxy fatty acids by column chromatography. The reaction is performed with 200 mg lipid and the supernatant after cent-

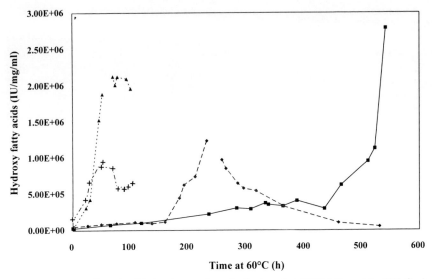

Figure 6.5 Progression of oxidation in a variety of oils stored at 60 °C followed by HPLC of hydroxy fatty acids derived from hydroperoxides. Peanut oil (■), chicken fat (♦), kiwi seed oil (▲) and fish oil (+).

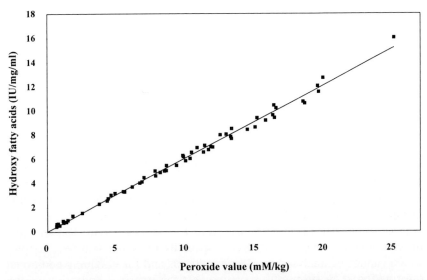

Figure 6.6 Correlation between hydroxy fatty acid measured by HPLC and peroxide value for canola oils. $y = 0.606x + 0$, $r = 0.997$, $n = 59$.

rifugation in a minimum of hexane is applied to a mini-column (15 ml) of Iatrobeads (1.5 g). The non-derivatised material is eluted with hexane–ether (10 ml, 5:95) and the hydroxy fatty acids with hexane–ether (10 ml, 50:50). The solvent is removed under nitrogen and the sample made up to

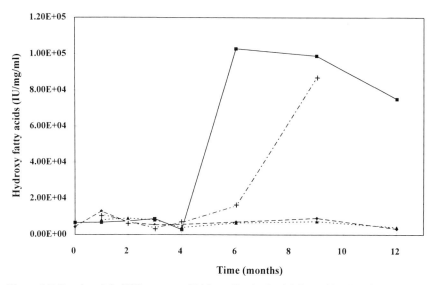

Figure 6.7 Results of shelf-life test at −25 °C on sliced salami followed by HPLC of hydroxy fatty acids. Salami treated with various antioxidants: non-stabilised (■), commercial rosemary extract, 50 ppm (◆), purified rosemary extract, 50 ppm (▲) and ternary antioxidant, 100 ppm (+).

200 µl for HPLC analysis as detailed above. This concentration technique is particularly useful for food samples with low levels of hydroperoxides, e.g. chocolate and milk powders.

Sensitivity of HPLC analysis can also be improved by using a more specific detector. A UV–vis detector at 234 nm, is very sensitive to conjugated dienes, but, as mentioned in section 6.4.1, not all hydroperoxides have conjugated diene groups, and measurement at 234 nm also detects other compounds, such as tocopherols. Electrochemical detectors have been used to detect selectively phospholipid hydroperoxides in biological tissues.[7] Hydroxy derivatives and non-oxidised phospholipids gave no response. Determination of arachidonic acid hydroperoxides and phospholipids in human plasma[8] and lipid hydroperoxides in the low nanogram range has also been reported.[9] A recent publication shows determination of hydroperoxides in several oils with a detection limit of 10.9 pmol for trilinoleoyl monohydroperoxides.[10] Correlations of 0.984 to 0.998 between PV and electrochemical detection were found.

Several researchers have reported the use of chemiluminescence detectors for assay of lipid hydroperoxides. Chemiluminescence is the emission of light resulting from radiative decay of an active intermediate in the excited state produced by chemical reactions. After separation by HPLC the hydroperoxides are detected by post-column reaction with luminol and a catalyst. One method based on the detection of chemiluminescence generated by the reaction of luminol with hydroeroxides and cytochrome c[11] had a high sensitivity to methyl linoleate hydroperoxide, arachidonic acid hydroperoxide and cholesterol hydroperoxide. Another system used the chemiluminescence emitted by isoluminol in the presence of hydroperoxide and microperoxidase.[12] It was applied to hydroperoxides from linoleic acid, phospholipids and cholesterol and was intended for biological rather than for food samples. More recently, determination of total hydroperoxides in corn oil has been reported[13] and also in solid matrices such as potato chips and French fried potatoes after fat extraction.

Fluorimetric detection of TG-OOHs after post-column reaction with diphenyl-1-pyrenylphosphine allowed their determination in the range 2–1000 pmol.[14] The same author found that this fluorimetric system gave an increased sensitivity more than 10 000 times that of conventional iodometry.[15]

All these HPLC techniques hold promise for more selective and sensitive detection of oxidation, especially at the low levels found in foods. A great advantage of using HPLC is that interfering substances, such as antioxidants which quench chemiluminescence, are separated before detection.

6.4.3 GC methods

Hydroperoxides can also be measured by GC, but require derivatization. The following method has been developed in our laboratories: the hydroperoxy group is reduced to an alchohol (with sodium borohydride). The methyl ester is formed by methylation with acetyl chloride by the following method: sample (1 mg) is heated at 100 °C for 1 h in the presence of methanol–hexane (4 ml, 4:1) and acetyl chloride (200 μl). The mixture is cooled and K_2CO_3 (5 ml, 6%) and hexane (7 ml) added. The upper layer is decanted and dried for silylation. A trimethylsilyl group is attached to the alcohol by reaction with BSTFA in acetonitrile (50 μl, 2:1) at room temperature for 1 h. The solvents are evaporated and the derivatives dissolved in hexane (1 ml) for separation on a CarboWax column and identification by mass spectrometry (MS). With this technique, hydroperoxides from C14:1 upwards have been found in milk chocolate. It was shown that although the fat extracted from the chocolate contained ca. 3% linoleic acid, ca. 65–68% of the total hydroperoxides came from linoleic acid (Table 6.3). Sensitivity was improved by purification of the hydroxy fatty acids by column chromatography as mentioned above (section 6.4.2). Hydroperoxides from a variety of oils were also identified. Sensitivity can also be increased by forming the pentafluorobenzyl esters instead of methyl esters and using an electrolytic conductivity detector.[16]

Table 6.3 Fatty acid composition and hydroperoxide composition in aerated chocolate[a]

	Fresh	8 months	14 months
	Fatty acid composition[b]		
16:0	30.5	29.4	33.7
18:0	28.8	29.7	29.5
18:1	32.0	30.9	27.2
18:2	3.3	3.1	2.9
	Hydroxy-TMS composition		
14:1	0.5	0.7	0.7
16:1	2.8	3.0	3.2
17:1	0.7	0.4	0.8
18:1	31.2	30.1	27.9
18:2	64.8	65.9	67.5

[a]Hydroxy-TMS represented ca. 90% of chromatogram. [b]Less than 3% not reported.

6.4.4 Indirect methods

Lipid hydroperoxides can also be measured indirectly by their breakdown products. Heating decomposes lipid peroxides to volatile components (e.g. pentane and hexanal) which can be quantified by headspace gas chromatography. For analysis of oils the sample is placed in a glass vial sealed with a crimp-cap septum. To generate volatiles, high temperatures (e.g. 140 °C for 30 min) are needed, and an inert atmosphere in the vials is essential for obtaining reproducible results (ca. 7% standard deviation).[17] To determine peroxides in dry foods, such as milk powder and soup mixes, we have modified a trap thermal desorber injector to determine the pentane produced by their heat-induced decomposition. The sample (300 mg) is mixed with an equivalent weight of glass beads and packed into a 10 cm glass tube. The tube is loaded into a TDAS 5000 automatic injector (Brechbühler, Geneva), the tube is purged to remove existing volatiles, pyrolysed at 110 °C for 30 s and the volatiles produced swept onto an alumina column (2 m, 3.2 mm ID packed with F-1 60/80 mesh) for analysis of the hydrocarbons by GC. Although this technique has a certain use for indicating oxidative quality of a sample after fabrication, it was difficult to use to follow the progression of oxidation during storage. Figure 6.8 shows the pentane produced by pyrolysis of various soup mixes manufactured at a range of water activities. The amounts of pentane correspond to initial oxidative quality of the products and indirectly to oxidative stability during storage. Pyrolysis has certain disadvantages: the sample tubes must be dried in a desiccator overnight to avoid transferring too much water onto the column; pyrolysis produces large amounts of dirt which contaminate the top of the column despite a

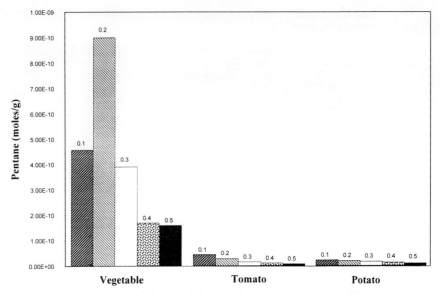

Figure 6.8 Pentane produced by pyrolysis of dehydrated soups (vegetable, tomato and potato) with water activities ranging from 0.1 to 0.5.

glass wool plug in the injector; and due to low reproducibility, multiple samples must be run. However, this method has the advantage of not requiring extraction of the lipid and of having, in our case, an automatic injector.

6.4.5 Other spectrometric methods

Chemiluminescence shows much promise as a sensitive, rapid measure of oxidation. It has been shown in our laboratories that the photon count in, for example milk powder, corresponds to sensory evaluation (Table 6.4) and it was also shown that addition of antioxidants decreased the production of photons.[18] However, it was necessary to heat the samples to at least 50° C to have a usable signal.

Table 6.4 Determination of chemiluminescence in milk powder[a]

Sample	Sensory	Count/10 s[b]
Milk powder	Fresh	2371
Milk powder	No longer fresh	3932
Milk powder	Slightly rancid	4503
Milk powder	Rancid	4910

[a]Heated at 50 °C. [b]± 10%.

There is interest in the determination of oxidation in oils by chemiluminescence from sodium hypochlorite-induced decomposition of hydroperoxides. This method has been reported for a variety of organic hydroperoxides and for lipid hydroperoxides in samples with a PV of less than 1[19] and more recently in autoxidized marine oils.[20] Although highly dependent on experimental conditions the technique is rapid, sensitive and uses small samples (100 µl). At the present time the equipment available is not sufficiently sensitive to the photons produced without some form of 'boost' (temperature or catalyst) as illustrated by the above results.

Recently, fluorimetrically-enhanced detection of hydroperoxides has been reported. Diphenyl-1-pyrenylphosphine (DPPP) is oxidised to strongly fluorescent DPPP oxide by hydroperoxides and the fluorescent intensity of DPPP oxide indicates the amounts of hydroperoxides present. This method was applied to oils and fatty foods (margarine and mayonnaise) and was found to be more than 10 000 times more sensitive than conventional iodometry.[15] The main drawback is that the reagent is not commercially available. Another method based on the oxidation of leucophloxin to fluorescent phloxin by hydroperoxide and haematin was adapted to stopped-flow injection.[21] The method was applied to lipohydroperoxides in oils and hydrogen peroxide in milk and showed good agreement with iodometric results. The presence of antioxidants decreased the fluorescence.

Under 'other methods' techniques such as NMR and IR should also be mentioned. In NMR the ratio of olefinic and divinylmethylene protons to aliphatic protons decreases during oxidation. This method has been applied to fish oils[22] and to oil extracted from salted dried fish.[23] Although it has the advantage of continuing to progress steadily after the PV has reached a maximum, it is too technologically expensive at present to be of use for routine determination of oxidation.

6.5 Secondary decomposition products

Although hydroperoxides are primary oxidation products they are tasteless and odourless and therefore do not contribute to the sensation of rancidity. They also have the disadvantage, from the analysts' view point, of being transient, as can be seen from the data presented above. Thus, samples with low levels of peroxides can be rancid. This is a more acute problem in foods compared to oils, where peroxides tend to accumulate past the point of rancidity. To give results which more closely correspond to sensory evaluation of oxidation, secondary decomposition products, for example aldehydes and ketones, can be measured.

6.5.1 Spectrophotometric methods

The thiobarbituric acid test (TBA) is one of the oldest and most commonly used methods for following lipid oxidation, especially in meat products. It involves reaction of TBA with aldehydes to produce a chromogen which is measured spectrophotometrically at 450 nm (alkanals) or 530 nm (alkadienals and alkenals) and is expressed in mg malonaldehyde/kg sample. The test can be performed directly on the food product, on an extract of the food, or on the steam distillate. Despite its wide application this method has been severely criticised.[24] Amongst these criticisms are:

- the exact nature of the reaction is still disputed
- malonaldehyde is a relatively minor lipid oxidation product formed from polyunsaturated fatty acids with three or more double bonds
- TBA can also react with other, non-lipid oxidation compounds to give red compounds.

Given these problems it has been recommended recently[25] that the TBA procedure be used as an overall indication of lipid oxidation rather than to quantify malonaldehyde. The term TBA value is now more commonly replaced by TBARS (thiobarbituric acid reactive substances). Despite the possibility of further improvement to the specificity, e.g. by HPLC separation of the complex, it is prudent to use other tests for lipid oxidation (e.g. hexanal levels) as a complement to the TBA test.

The anisidine value (AV) is another widely used test and is empirically defined as 100 times the absorbance of a solution from the reaction of 1 g oil in 100 ml solvent and *p*-anisidine measured at 350 nm in a 10 mm cell. The *p*-anisidine reagent forms a condensation product with conjugated dienals and alka-2-enals.[26] It is of particular use for abused oils with low PV, e.g. frying oils and oils after refining, but it has also been used in foods. For example, the effect of product composition (e.g. addition of metals and variation of water activity) and storage conditions on oxidative stability of spray-dried milk powder has been followed by measuring the AV of the free fat extract.[27]

Aldehydes can also be measured by reaction with N,N-dimethyl-*p*-phenylenediamine (DPPD) in the presence of acetic acid with spectrophotometric detection at 400, 460 and 500 nm for alkanals, alkenals and alkadienals, respectively.[28] This technique has the advantage of being simpler and more rapid than the AV. The solvent benzene has been replaced with toluene to reduce toxicity. Development of AV and DPPD in blackcurrant seed oil containing a number of antioxidants is shown in Table 6.5.

Carbonyls can be measured as their 2,4-dinitrophenylhydrazones at 440 nm after reaction with 2,4-dinitrophenyl hydrazine (2,4-DNPH). This classic technique is also one of the most widely used. However, it has the disadvantage that hydroperoxides decompose under the experimental conditions and give falsely high values. Several solutions have been proposed,[29] among the more elegant of which is to reduce the hydroperoxides with triphenylphosphine before reaction.[30] Individual 2,4-dinitrophenylhydrazones can be separated by HPLC and GC (see sections 6.5.2 and 6.5.3.4 below).

All the above methods can be easily applied to fats and oils. However, we have found losses of hexanal of up to 75% during solvent evaporation (hexane) after extraction from a food, which seriously decreases the sensitivity and as a consequence the usefulness of these methods. *In situ* derivatisation followed by extraction would be more efficient.

6.5.2 HPLC methods

HPLC techniques are most often applied to biological systems. Measurement of aldehydic lipid peroxidation products as 2,4-dinitrophenylhydrazones, as fluorescent decahydroacridine derivatives after reaction with 1,3-cyclohexanedione, and analysis of free 4-hydroxyalkenals and malonaldehyde for these applications have been reviewed.[31] Aldehydes can also be measured by post-column fluorimetric detection after reaction with 7-hydrazino-4-nitrobenzo-2,1,3-oxadiazole (NBD hydrazine) to form NBD hydrazones.[32]

Reaction with 2,4-DNPH is the most commonly used method. The hydrazones are stable and non-volatile, which is an advantage for sample preparation. However, HPLC chromatograms obtained can be complicated and difficult to resolve. Prior separation of the different classes by thin layer chromatography (TLC) is useful, but the technique becomes too long for routine use. In non-biological systems 2,4-DNPH has been extensively applied for analysis of the environment (water and in particular air and exhaust gases). Application directly in foods has been limited to whisky,[33] meat,[34] shrimp,[35] milk powder[36] and soyabean powder.[37] The major problem is contamination of the reagents and solvents by aldehydes (beware smokers!) which necessitates purification. Its application can be enlarged by isolation of the volatiles by for example distillation into a trap filled with 2,4-DNPH. This technique has been applied to meat[38] and fried fish.[39]

6.5.3 GC methods

Analysis by GC is the most powerful tool available for determining volatile oxidation products due to its high resolution and the possibility of coupling a mass spectrometer for compound identification. The products can be analysed directly (e.g. by headspace or simultaneous distillation extraction (SDE) extract) or as more stable derivatives. Analysis of

Table 6.5 Comparison of determination of carbonyl content in oils by anisidine value and DPPH

Time (weeks)	Anisidine value						DPPH ($\mu g\,mg^{-1}$)				
	Non-stabilised	Ascorbyl palmitate	Nipanox	Ternary antioxidant	Quaternary antioxidant		Non-stabilised	Ascorbyl palmitate	Nipanox	Ternary antioxidant	Quaternary antioxidant
0	20	19	18	19	17		0	0.6	0.6	0.3	0
3	152	183	–[a]	–	–		6.8	13.2	–	–	–
6	–	–	144	106	–		–	–	7.1	4.4	–
8	376	392	–	–	37		16.7	18.3	–	–	1.3
10	–	–	381	–	–		–	–	18.8	–	–
14	–	–	–	247	353		–	–	–	10.0	15.4

[a]Not determined

volatile oxidation products from oils and fats and foods implies isolation of these compounds. There are a number of extensive reviews of different GC-based methods available.[40,41] Several techniques: static headspace, purge and trap, and SDE are discussed below.

6.5.3.1 Static headspace. This involves direct injection of the headspace gases above a food sample in a sealed container onto a gas chromatograph for analysis. Headspace sampling most accurately represents the flavour/odour above a food, but sensitivity is a problem. The amount of headspace that can be injected onto the column is restricted to ca. 5 ml or less, especially for capillary columns. Larger volumes result in poor resolution and bad chromatography, in particular for very volatile compounds. The minimum amount of sample that can be detected corresponds to ca. 0.1 ppm in the headspace, and for MS ca. 10 ppm is required. Since the concentration of volatiles in the headspace ranges from 0.01 ppb to 100 ppm, only the most abundant can be analysed and thus this technique is not suitable for trace compounds.[41] However, direct headspace sampling has the advantages of simplicity and rapidity and has been widely used. In our laboratories we routinely analyse hydrocarbons (ethane and pentane) together with residual oxygen (see section 6.2) for following oxidation in a wide range of food products.[42] These analyses were originally preformed on an alumina column (2 m, 3.2 mm ID, packed with F-1 60/80 mesh), but the sensitivity and peak resolution can be improved with a megabore GS-alumina column (30 m, 0.53 mm ID; J & W, Folsom, CA). Decrease in residual oxygen and development of ethane in the headspace of oat-based cereal product treated with various antioxidants is shown in Figure 6.9, to illustrate the use of this method.

Direct headspace sampling is particularly useful for very volatile compounds, since other techniques which require the container to be opened and a sample taken lose a significant proportion of these compounds during the transfer. To determine aldehydes and ketones we use an automatic headspace sampler (HS40, Perkin-Elmer) with an SE-54 capillary column (30 m, 0.32 mm ID, 4 μm film thickness) and a temperature program from 5 to 250 °C at 5 °C min^{-1}. These conditions allow the separation of ca. 50 lipid oxidation volatiles; also, injection at 5 °C avoids the formation of ice plugs in the column, which occurs at the temperatures below zero which are necessary with other columns. Certain advantages and constraints of this sytems should be noted. Samples must be conditioned in glass vials (max. 22 ml) which seriously limits sample size (up to ca. 5 g) and presents problems with sample homogeneity. The sample must be thermostatted prior to injection, which forces more volatiles into the headspace and so increases sensitivity. However, it has been found that heating to more than 80 °C for 30 min causes decomposition of the volatiles. The pressurisation injection system allows only ca.

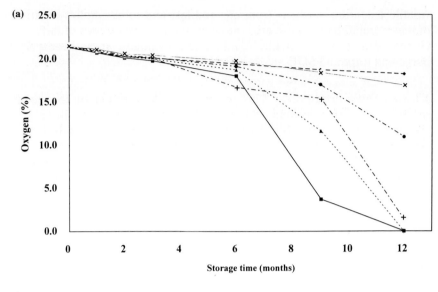

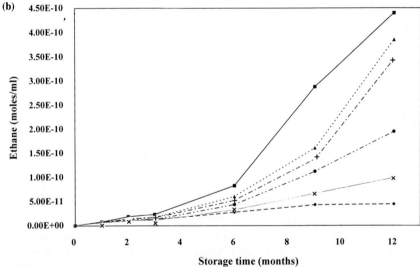

Figure 6.9 Results of shelf-life test at 30 °C on oat-based cereal: non-stabilised (■), BHA, 100 ppm (♦), propyl gallate, 100 ppm (▲), rosemary extract, 200 ppm (+), 400 ppm (●) and 800 ppm (×). (a) Residual oxygen and (b) headspace ethane.

1 ml to be injected without reducing column efficiency, which reduces the sensitivity but avoids carry-over from syringe injection. Since the headspace is injected the concentration of a particular compound is limited by its partition coefficient and its boiling point; the upper limit has been found to be around 2,4-heptadienal (b.pt. 95 °C). The system is especially

useful for relatively concentrated headspace samples, e.g. headspace above oils and oily foods, and for highly volatile compounds.

This system can be used to follow the oxidation of foods during a shelf-life study, but another approach is to use it as a measure of oxidation status at a particular moment. For this, a sample is conditioned in a vial capped with a septum and thermostatted, to drive the volatiles present into the headspace which is subsequently analysed by GC (e.g. for corn chips[43] and cooked chicken meat[44]). This technique resembles the indirect methods for measuring peroxides outlined in section 6.4.4 above and has also been used for determining volatiles from oils produced by thermolysis (e.g. at 180 °C for 20 min[45]).

A further modification is the addition of water, more particularly of solutions of salts, to force the volatiles into the headspace. However, this can lead to problems of water interference on the column and also the concentration of some volatiles can even be reduced. It should also be remembered that the headspace is no longer the same as originally present since all volatiles do not behave in the same way.

6.5.3.2 Purge and trap. The problems of low sensitivity encountered with direct headspace sampling can be overcome by purge and trap techniques. In these the volatiles are purged from the sample by a stream of inert gas (helium or nitrogen) and are trapped on a cartridge of for example Tenax, activated charcoal or Chromasorb. The volatiles are desorbed from the trap by rapid heating and are analysed by GC. This technique is also known as dynamic headspace, and is one of the most widely used methods. Tekmar is a well-known commercial system. The major advantage of purge and trap is that volatiles from large amounts of sample can be concentrated which increases sensitivity, especially for trace analysis. The disadvantages include break-through, especially of the very volatile compounds, alteration or even destruction of compounds during thermal desorption, low recovery of high boiling point compounds and transfer of water to the analytical column (especially from high moisture foods). The method is also sensitive to purge, trap and desorption conditions which must be rigorously controlled.[46] It can be used to follow oxidation in foods although it is more frequently used for flavour studies. In a study on oxidation of salmon during frozen storage it has been found to be complementary to SDE (see section 6.5.3.3 below) in that purge and trap determines the more volatile substances whereas SDE extracts and determines the less volatile substances (J.-C. Spadone, 1993, personal communication).

The combination of extraction with supercritical carbon dioxide and trapping of the volatiles entrained in the gas stream when the pressure is released on a cartridge for subsequent analysis by desorption GC-MS is a promising development (U. Stollman, 1993, personal communication).

6.5.3.3 Simultaneous distillation extraction (SDE). This is also known as Nickersen–Likens extraction and involves distillation of an aqueous solution of the food and simultaneous extraction of the distillate with solvent. Typically, the solvent is dried over sodium sulphate and the volume is reduced to increase the concentration appropriately before injection onto a gas chromatograph for analysis. This is not a gentle technique as it involves boiling the sample and volatiles are also lost during evaporation of the solvent. The technique can be improved by using a micro-SDE which gives a more concentrated solution and avoids evaporation, or performing under vacuum which reduces the severity of heat treatment. SDE is extensively used for extraction of flavours but has been less widely applied to determination of oxidation in foods. However, coupling of the gas chromatograph to a mass spectrometer which allows identification of the compounds makes this a very powerful and potentially useful technique. For example, in a recent study of dietetic milk powders, 84 compounds were identified by GC–MS; aldehydes and in particular hexanal were quantitatively the most important (D. Beauverd and W. Budnik, 1991, personal communication).

Advantages of SDE compared to headspace techniques include extraction of higher-boiling point substances with reduced volatility and greater concentrations, allowing identification by MS and the possibility of simultaneous sniffing of the element. Disadvantages include destruction of volatiles during distillation, loss of volatiles during concentration and 'loss' of volatiles in the solvent peak during analysis.

6.5.3.4 Derivatives. As mentioned above (sections 6.5.1 and 6.5.2), carbonyl compounds can be derivatised to their 2,4-DNP hydrazones, and although usually determined by HPLC, they can also be analysed by GC. Under GC analysis conditions the 2,4-DNP hydrazones form *syn* and *anti* isomers: on the one hand this complicates the chromatogram, but on the other hand it can facilitate identification.[47] Coupling of GC to MS allows positive identification of the compounds.[48]

Carbonyl compounds can also be derivatised with cysteamine, morpholine or *N*-methylhydrazine.[49] These methods have been most commonly applied by purging the headspace from, for example heated cooking oils, into a trapping solution containing the derivatising agent. However, the reaction with cysteamine has also been applied to foods in solution, e.g. coffee, milk powder, wine and fruit juices.[50] To improve sensitivity a nitrogen–phosphorus sensitive detector was used.

6.6 Fluorescent products

There has been growing interest in determination of oxidation products by fluorescence techniques due to their potentially high sensitivity.

Fluorescent compounds were first measured in biological samples[29] and the technique was found to be ca. 10–100 times more sensitive than the colorimetric TBA assay for measuring Schiff base products.[51] There is less literature available for analysis of food oxidation. In the first application published, fluorescence in solvent extracts from freeze-dried meats was demonstrated.[52] More recently, applications of extractable fluorescence for following oxidation in salted sun-dried fish[53] and processed pork meat[54] have been shown. In the authors' laboratories fluorescence of the organic phase after Folch extraction with chloroform–methanol has been used to detect oxidation in frozen salami (Figure 6.10). This approach is relatively simple and rapid, but is not specific, since the identity of the compounds is not known. Separation by TLC or HPLC would be helpful. Separation of β-lactoglobulin after reaction with aldehydes by size exclusion chromatography and measurement of the fluorescence is a promising approach (L. Skibsted, 1993, personal communication).

Measurement of fluorescence in the solid state has also been reported. In one technique chitosan powder was exposed to volatiles from oxidising lipids and its subsequent fluorescence measured in the solid state;[55] Weist and Karel concluded that the fluorescence was due to volatile malonaldehyde. In another approach solid sample fluorescence was used for assessment of oxidation in freeze-dried fish with the aim of measuring protein-bound fluorescence which was insoluble in organic solvents.[56]

In conclusion, lipid oxidation, especially in foods, is a highly complex

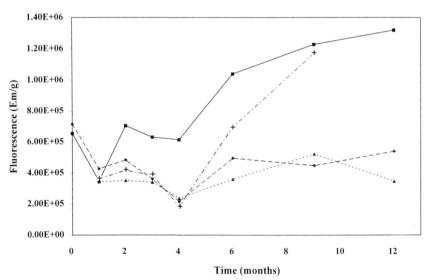

Figure 6.10 Development of organic phase fluorescence in sliced salami stored at −25 °C. Salami treated with various antioxidants: non-stabilised (■), commercial rosemary extract, 50 ppm (♦), purified rosemary extract, 50 ppm (▲) and ternary antioxidant, 100 ppm (+).

process resulting in many and varied products. At present no universal technique exists for determining oxidation and the food scientist must use a number of methods to give an overall indication. In this chapter we have covered the most important, and the most promising spectrometric and chromatographic methods currently available for determining oxidation.

References

1. Kubow, S. (1993). *Nutr. Rev.*, **51**, 33.
2. Belitz, H.-D. and Grosch, W. (1992). *Lehrbuch der Libensmittelchemie*, Springer-Verlag, Berlin, p. 186.
3. Windle, J. J., Nimmo, C. C. and Lew, J.-L. (1976). *Cereal Chem.*, **53**, 671.
4. Parr, L. J. and Swoboda, P. A. T. (1976). *J. Food Technol.*, **11**, 1.
5. Hills, G. L. and Thiel, C. C. (1946). *J. Dairy Res.*, **14**, 340.
6. Park, D. K., Terao, J. and Matsushita, S. (1981). *Agric. Biol. Chem.*, **45**, 2443.
7. Yamada, K., Terao, J. and Matsushita, S. (1987). *Lipids*, **22**, 125.
8. Terao, J., Shibata, S. S., Yamada, K. and Matsushita, S. (1989). In *Medical Biochemical and Chemical Aspects of Free Radicals*, (eds O. Hayaishi, E. Niki, M. Kondo and T. Yoshikawa), Elsevier Science Publishers, Amsterdam, pp. 781–788.
9. Funk, M. O. (1987). *Free Radical Biol. Med.*, **3**, 319.
10. Song, J. H., Chang, C. O. and Park, D. K. (1992). *Korean Biochem. J.*, **25**, 337.
11. Miyazawa, T., Fujimoto, K. and Kaneda, T. (1987). *Agric. Biol. Chem*, **51**, 2569.
12. Yamamoto, Y., Brodsky, M. H., Baker, J. C. and Ames, B. N. (1987). *Anal. Biochem.*, **160**, 7.
13. Yang, G. C. (1992). *Trends in Food Sci. Technol.*, **3**, 15.
14. Akasaka, K., Ijichi, S., Watanabe, K., Ohrui, H. and Meguro, H. (1992). *J. Chromatogr.*, **596**, 197.
15. Akasaka, K., Sasaki, I., Ohrui, H. and Meguro, H. (1992). *Biosci. Biotech. Biochem.*, **56**, 605.
16. van Kuijk, F. J. G. M., Thomas, D. W., Stephens, R. J. and Dratz, E. A. (1983). *J. Free Radicals Biol. Med.*, **1**, 387.
17. Ulberth, F. and Roubicek, D. (1993). *Food Chem.*, **46**, 137.
18. Löliger, J. and Saucy, F. (1984). *J. Lumin.*, **31 & 32**, 908.
19. Yamamoto, Y., Niki, E., Tanimura, R. and Kamiya, Y. (1985). *J. Am. Oil Chem. Soc.*, **62**, 1248.
20. Burkow, I. C., Moen, P. and Øverbø, K. (1992). *J. Am. Oil Chem. Soc.*, **69**, 1108.
21. Pérez-Ruiz, T., Martínez-Lozano, C., Tomás, V. and Val, O. (1993). *Food Chem.*, **46**, 301.
22. Saito, H. and Nakamura, K. (1990). *Agric. Biol. Chem.*, **54**, 533.
23. Saito, H. and Udagawa, M. (1992). *J. Am. Oil Chem. Soc.*, **69**, 1157.
24. Ward, D. D. (1985). *Milchwissenschaft*, **40**, 583.
25. Gray, J. I. and Monahan, F. J. (1992). *Trends in Food Sci. Technol.*, **3**, 315.
26. Robards, K., Kerr, A. F. and Patsalides, E. (1988). *Analyst*, **113**, 213.
27. Roozen, J. P. and Linssen, J. P. H. (1992). In *Lipid Oxidation in Food*, (ed. A. J. St. Angelo), American Chemical Society, Washington, DC, pp. 302–309.
28. Miyashita, K., Kanda, K. and Takagi, T. (1991). *J. Am. Oil Chem. Soc.*, **68**, 748.
29. Gray, J. I. (1978). *J. Am Oil Chem. Soc.*, **55**, 539.
30. Chiba, T., Takazawa, M. and Fujimoto, K. (1989). *J. Am. Oil Chem. Soc.*, **66**, 1588.
31. Esterbauer, H. and Zollner, H. (1989). *Free Radical Biol. Med.*, **7**, 197.
32. Koizumi, H. and Suzuki, Y. (1988). *J. Chromatogr.*, **457**, 299.
33. Puputti, E. and Lehtonen, P. (1986). *J. Chromatogr.*, **353**, 163.
34. Melton, S. L. (1983). *Food Technol.*, **37**, 105.
35. Radford, T. and Dalsis, D. E. (1982). *J. Agric. Food Chem.*, **30**, 600.

36. Buckley, K. E., Fisher, L. J. and MacKay, V. G. (1986). *J. Assoc. Off. Anal. Chem.*, **69**, 655.
37. Tran, Q. K., Takamura, H. and Kito, M. (1992). *Biosci. Biotech. Biochem.*, **56**, 519.
38. Reindl, B. and Stan, H.-J. (1982). *J. Agric. Food Chem.*, **30**, 849.
39. Lane, R. H. and Smathers, J. L. (1991). *J. Assoc. Off. Anal. Chem.*, **74**, 957.
40. Waltking, A. E. and Goetz, A. G. (1986). *CRC Crit. Rev. Food Sci. Technol.*, **19**, 99.
41. Heath, H. B. and Reineccius, G. (1986). *Flavor Chemistry and Technology*, Macmillan, Basingstoke, pp. 1–42.
42. Löliger, J. (1990). *J. Sci. Food Agric.*, **52**, 119.
43. Robards, K., Kerr, A. F., Patsalides, E. and Korth, J. (1988). *J. Am. Oil Chem. Soc.*, **65**, 1621.
44. Ang, C. Y. W. and Young, L. L. (1989). *J. Assoc. Off. Anal. Chem.*, **72**, 277.
45. Snyder, J. M., Frankel, E. N., Selke, E. and Warner, K. (1988). *J. Am. Oil Chem. Soc.*, **65**, 1617.
46. Vercellotti, J. R., Mills, O. E., Bett, K. L. and Sullen, D. L. (1992). In *Lipid Oxidation in Food*, (ed. A. J. St. Angelo), American Chemical Society, Washington, DC, pp. 232–265.
47. Uralets, V. P., Rijks, J. A. and Leclercq, P. A. (1980). *J. Chromatogr.*, **194**, 135.
48. Stanley, J. B., Brown, D. F., Senn, V. J. and Dollear, F. G. (1975). *J. Food Sci.*, **40**, 1134.
49. Shibamoto, T. (1989). In *Flavors and Off-flavors*, (ed. G. Charalambous), Elsevier, Amsterdam, pp. 471–484.
50. Hayashi, T., Reece, C. A. and Shibamoto, T. (1985). In *Characterization and Measurement of Flavor Compounds*, (eds D. D. Bills and C. J. Mussinan), American Chemical Society, Washington, DC, pp. 61–78.
51. Dillard, C. J. and Tappel, A. L. (1971). *Lipids*, **6**, 715.
52. Kamerei, A. R. and Karel, M. (1984). *J. Food Sci.*, **49**, 1517.
53. Smith, G. and Hole, M. (1991). *J. Sci. Food Agric.*, **55**, 291.
54. Madsen, H. L., Stapelfeldt, H., Bertelsen, G. and Skibsted, L. H. (1993). *Food Chem.*, **46**, 265.
55. Weist, J. L. and Karel, M. (1992). *J. Agric. Food Chem.*, **40**, 1158.
56. Hasegawa, K., Endo, Y. and Fujimoto, K. (1992). *J. Food Sci.*, **57**, 1123.

7 Nutritional aspects of rancidity
T. A. B. SANDERS

7.1 Introduction

The word 'rancid' is derived from *rancidus*, the Latin for 'stinking'. The term 'rancidity' is used to describe the taste or smell of rank stale fat. Rancid fat contains a wide variety of chemical substances, whose structures are not all known. Human taste buds are highly sensitive to some compounds such as lactones and free fatty acids, so only minute amounts of these compounds need to be produced to spoil the taste of food. Rancidity is undesirable because it leads to waste and a deterioration in quality. Hydrolytic rancidity, which is caused by the release of free fatty acids from glycerides, is extremely important in determining how a product tastes but is unlikely to be of any toxicological significance because fats are in any case hydrolysed enzymatically in the small bowel before they are absorbed. In some cases hydrolytic rancidity is regarded as desirable: for instance, in some strong cheeses such as Stilton, where it gives the cheese a sharp burning taste. Oxidative rancidity, on the other hand, leads to the formation of both unpalatable and toxic components and destroys nutrients.

7.2 Toxic components

There are several classes of material in oxidised fat that have toxic effects:[1-3] peroxide fatty acids and their subsequent end-products, polymeric material and oxidised sterols.
The toxic components are:

- Lipid peroxides
- Hydroxy fatty acids
- Carbonyl compounds—malonaldehyde
- Cyclic monomers
- Dimers and polymers
- Polycyclic aromatic hydrocarbons
- Oxidised sterols

The pharmacological activity of breakdown products of peroxidised fatty acids varies but may be very potent because they resemble biochemical messengers derived enzymically from polyunsaturated fatty acids in the body.[4] Polymeric material is poorly digested but may interact with gut contents and affect absorption of fat-soluble vitamins. Oxidised sterols, particularly cholesterol, may possess hormone-like activity. Other potentially toxic material results from the pyrolysis of fats on grilling and roasting meat and fish, and includes polycyclic aromatic hydrocarbons, which are known carcinogens.

Oxidative rancidity also decreases the nutritional quality of food. Lipid peroxides and the free radicals they generate destroy fat-soluble vitamins A and E in food and react with sulphydryl bonds in proteins. As sulphur amino acids are often the nutritionally limiting amino acids in many proteins, a reduction in the sulphur amino acid content will invariably lead to a decrease in protein quality. This probably explains why the protein quality of smoked mackerel is lower that that of fresh mackerel.[5]

7.3 Dietary sources of oxidised fats

Oxidative rancidity occurs under conditions used in the preparation, processing and storage of foods. Virtually every type of food has been reported to contain lipid peroxides.[2] The oxidative process is accelerated by heat, light, water and certain metal ions, notably iron and copper. The susceptibility of different fats to oxidative rancidity varies, depending upon the degree of unsaturation of the component fatty acids, the presence of transition metals such as iron and copper and availability of antioxidants. Products with a large surface area are potentially more at risk from oxidative rancidity. Foods that are of most concern are dehydrated products, fish products and frying oils.

Eggs are the most significant source of dietary cholesterol in human diets and it has been suggested that dried eggs contain significant amounts of oxidised cholesterol. Offal and shellfish are other foods relatively rich in cholesterol and related sterols. Vegetable oils contain other sterols such as β-sitosterol and these can also undergo oxidative changes during deep-fat frying.

Polyunsaturated fatty acids are more susceptible to oxidation than monounsaturated fatty acids. Moreover, the greater the number of double bonds in the fatty acid, the more prone it is to peroxidation. Foods, particularly those of vegetable origin, contain linoleic (18:2, $n-6$) and α-linoleic (18:3, $n-3$) acids. γ-Linolenic acid (18:3, $n-6$) is found in certain oils such as borage, evening primrose and blackcurrant seed oils. Linoleic acid is the predominant polyunsaturated fatty acid in most vegetable oils such as sunflower, cotton seed, soyabean and corn oils (Figure 1: see also chapter 1).

α-Linolenic acid is far more prone to oxidation than linoleic acid, and this property has been employed for many years by the varnish and paint industry in its use of linseed oil, which contains about 60% α-linolenic acid, as a drying oil. With the exception of soyabean and rapeseed oil, most culinary vegetable oils only contain small amounts of α-linolenic acid. The oxidation of α-linolenic acid leads to fishy taints. For this reason soyabean and rapeseed oils are often selectively hydrogenated to decrease the concentration of α-linolenic acid. From a nutritional point of view this may be undesirable because α-linolenic acid may be essential in the diet.

Oxidative rancidity of refined vegetable oils is generally not a major problem for two reasons: linoleic acid is invariably the major polyunsaturated fatty acid, and is generally accompanied by considerable amounts of naturally occurring antioxidants such as vitamin E as well as synthetic antioxidants. It has, however, been claimed that unrefined 'coldpressed oils' deteriorate more rapidly than conventionally refined oils that contain synthetic antioxidants.[6] Oxidation of vegetable oils on storage is not particularly rapid. Potato crisps used to have antioxidants added to prevent rancidity that occurred on storage, but since the introduction of 'sell-by-dates' this is no longer found to be necessary.

Animal fats are more prone to oxidation because they are low in vitamin E, their polyunsaturated fatty acids are more unsaturated and in

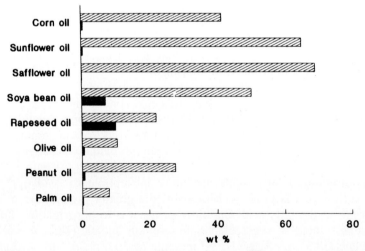

Figure 7.1 Proportions of linoleic (▨) and α-linolenic (■) acid in vegetable oils.

meat the iron in haemoglobin and myoglobin accelerates lipid peroxidation. Fish lipids are even more susceptible to oxidation and are probably responsible for the rapid spoilage of oily fish such as mackerel and herring.

The polyunsaturated fatty acid content of animal fats depends upon the animals' dietary intake of fat, with the notable exception of ruminants (sheep, cows, goats), which always have a low content. Thus pigs fed on corn oil will deposit fat high in linoleic acid. Animals do not usually synthesise dietary polyunsaturates, but they can convert linoleic and α-linolenic acids into more unsaturated derivatives with up to six double bonds with a carbon chain length of 20–22, notably arachidonic acid (20:4, $n-6$), eicosapentaenoic acid (20:5, $n-3$) and docosahexaenoic acid (22:6, $n-3$). These C_{20-22} polyunsaturated fatty acids are found chiefly in membrane phospholipids and are highly susceptible to peroxidation. It is the oxidation of these fatty acids that lead to off-taints in meat. The membrane lipids in meat are particularly susceptible to oxidation because it has a relatively high iron content.

The highest concentrations of C_{20-22} polyunsaturated fatty acids are found in fish and fish oils; many fish oils contain about 20% of their total fatty acids as these higher polyunsaturated fatty acids. This explains the susceptibility of fish to oxidative rancidity.[7] On oxidation, it appears that eicosapentaenoic acid and docosahexaenoic acid rapidly polymerise rather than form stable peroxides or hydroperoxides. Consequently, peroxide values are unreliable indicators of lipid peroxidation in fish oils.[8] Smoked oily fish such as kippers and smoked mackerel also contain significant amounts of oxidised lipids. Sun-dried oily fish such as anchovy are consumed in many parts of the world; for example 'ikan bilis' in Indonesia and Malaysia. In these products the polymerised material forms a protective coating on the outside of the fish and stabilises it from further oxidation.[9]

Although large amounts of fish oils are used in margarine manufacture in the UK, Norway and Germany, they are partially hydrogenated before use. This process converts the C_{20-22} polyunsaturated fatty acids into a mixture of positional and geometric isomers of 20:1, 20:2, 22:1 and 22:2.[10] Partially hardened fish oil is one of the most oxidatively stable fats and is widely used in the bakery industry.

Cod-liver oil is one of the few fish oils to be consumed without prior hardening. Most cod-liver oils have antioxidants added to protect against autoxidation. More recently, fish oil concentrates have been marketed for the treatment of hyperlipidaemia. In the USA such preparations are widely sold in health food shops for the prevention of coronary heart disease. However, many of these do not contain synthetic antioxidants because of consumer pressure to remove food additives.

7.4 Food processes that lead to the oxidation of fats

Cooking leads to lipid oxidation. The amount of oxidised material formed depends upon temperature, the availability of oxygen and duration of exposure. Malonaldehyde (MDA) concentrations have been used as a measure of lipid peroxidation in foods during cooking.[11,12] Boiling or microwave cooking leads to the lowest increments in MDA concentrations, whereas frying or baking leads to large increases in MDA concentrations in meat. In contrast, MDA in fish tends to decrease upon cooking (Figure 7.2).

When meat and fish are exposed to the intense heat of the flame in a charcoal broiler, and the fat is pyrolysed, polycyclic aromatic hydrocarbons (PAH) are formed. Smaller amounts of PAH are formed during roasting and deep frying. The concentration of PAH can reach 200 $\mu g\,kg^{-1}$ during charcoal broiling. The concentration of PAH in food can be reduced by altered processing and cooking methods.[5] For example, charcoal broiling and lower temperatures, with reduced fat content in the meat, and avoidance of direct contact of the meat and fat with the flame.

7.4.1 Thermally oxidised fats

Deep-fat frying is normally carried out at a temperature of 180 °C. It is also necessary to take into account any specific interaction the food may have with the oil and the conditions under which the oil is used. The

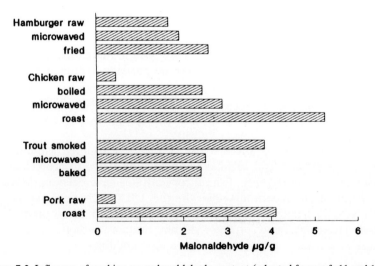

Figure 7.2 Influence of cooking on malonaldehyde content (adapted from refs 11 and 12).

evolution of water vapour from food results in the steam distillation of the volatiles from the frying fat, including vitamin E and antioxidants such as butylated hydroxytoluene (BHT). Heating fats and oils results in oxidation and the formation of 'polar material', which includes partial glycerides and free fatty acids.[13] Shallow-pan frying is accomplished quickly and repeated re-use of the fat is not common practice. Consequently the proportion of polar material formed is not high (about 10%). On the other hand, with deep-fat frying, the fat may be re-used and may be kept hot for extended periods. In commercial deep-fat frying, re-use is dictated by economics. The proportion of polar material can reach over 25% after 30 h deep-fat frying. Oils containing more than 30% polar material becomes unacceptable for cooking owing to changes in taste and odour.[13] Industrial deep frying yields similar amounts of polar material to shallow frying (about 10%) because the rate of turnover of the fat is high.

The polar material in thermally oxidised fat contains small amounts of cyclic monomers ranging from 0.1 to 0.5% by weight.[14] These cyclic monomers are formed mainly from linoleic and linolenic acids, and are known to be toxic. They were originally isolated from linseed oil that had been heated at 275 °C in the absence of oxygen. They were isolated as the distillable non-urea adductable fraction.

7.4.2 Food irradiation

Saturated and monounsaturated fatty acids are very resistant to ionising radiation but polyunsaturated fatty acids with two or more double bonds are highly vulnerable. Wills[15] found minimal formation of lipid peroxides in artificial food mixtures that consisted of lard:starch or corn oil:starch mixtures, but found that very large concentrations of peroxides were formed in irradiated herring oil:starch mixtures. The susceptibility of herring oil to lipid peroxidation was attributed to its high content of C_{20-22} polyunsaturated fatty acids. Peroxide yields were greater if the fat was dispersed in an inert medium such as starch than when irradiated in the pure form, and were also dependent on the presence of water in the dispersant medium. Proteins such as casein were found to inhibit peroxide formation. Tyrosine and glycine were found to decrease peroxide formation by about 30%.

These findings would suggest that irradiation of oily fish could lead to a marked build-up of lipid peroxides. Although food irradiation is not practised in the UK, it is in other parts of the EU. There is considerable commercial pressure to introduce food irradiation for poultry in order to decrease food poisoning caused by *Salmonella*. However, poultry feeds often contain significant amounts of fish oil provided as fish meal. Consequently, irradiation of poultry could lead to the generation of lipid peroxides.

7.5 Biochemical effects resulting from lipid peroxides

One of the first changes that occurs on lipid peroxidation is a decrease in C_{20-22} polyunsaturated fatty acids in membrane phospholipids, leading to breakdown of cell membranes and the release of membrane components.[16] Lipid hydroperoxides may activate or inhibit certain enzymes. They may lead to the oxidation of thiol groups in proteins leading to altered protein conformation. The breakdown products will then spread to other tissues to produce damaging effects.

The paradox of why living matter does not go rancid until it in turn becomes food can be explained by the existence of metabolic systems that mop-up free radicals, e.g. the enzymes superoxide dismutase, glutathione peroxidase, catalase and the antioxidant vitamins E and C.[17] This protective system can be overloaded by a high intake of lipid peroxides. The capacity to cope with lipid peroxides is dependent upon adequate intakes of selenium, copper, zinc, sulphur amino acids and vitamins C and E. A high intake of polyunsaturated fat can also exacerbate an underlying dietary deficiency of these protective nutrients. For example, cattle or sheep which have marginal intakes of selenium when fed new grass in spring develop a form of muscular dystrophy called white muscle disease. The same symptoms can be induced by feeding the animals polyunsaturated fats in their ration. Linoleic, eicosapentaenoic and docosahexaenoic acids are more effective than α-linolenic acid in exacerbating this disorder.[18]

The capacity to cope with free radicals may also be compromised by tissue injury or disease. Tissue injury often results in free radical formation and this is why free radicals have been associated with several diseases, for example arthritis, cancer and atherosclerosis.[19] The fact that increased free radical reactions can be demonstrated in almost any disease does not necessarily imply a causal relationship. However, a failure to control the production of free radicals promotes tissue injury. This is well illustrated by the toxic liver injury caused by alcohol or by chemicals such as carbon tetrachloride. Lipid peroxidation *in vivo* can be measured by pentane and ethane production, which can be determined on expired air,[20] or by measuring the thiobarbituric acid–malonaldehyde complex[16] in urine.

7.6 Acute effects

Diarrhoea is a common symptom and is probably caused by the formation of fatty acids with hydroxy groups from fatty acid hydroperoxides. The secondary lipid oxidation products are mainly responsible for the acute adverse biological effects resulting from the consumption of rancid fats.

Tissue congestion, fatty degeneration and necrosis were more severe in mice dosed with autoxidised methyl linoleate containing secondary oxidation products than with methyl linoleate hydroperoxides.[3]

The unpleasant taste of rancid fat is one safeguard against it being consumed: fats with a peroxide value of greater than 10 tastes rancid. However, not all individuals are able to control what they eat and infants receiving milk formulae and patients receiving nasogastric feeds are potentially at risk.

7.7 Chronic effects

McCollum[21] reviewed the early work on the toxicity of rancid fat. In 1923 Hopkins reported that diets containing a high fat content provided by cod-liver oil caused injury to rats. In 1926 Agdurh showed that small amounts of rancid cod-liver oil given continuously produced injurious effects, especially on the heart muscle, in mice, rats, rabbits, dogs, calves, cats, pigs and human infants. In describing the lesions he noted fatty degeneration of the heart muscle and replacement of the muscle fibre by connective tissue. Whipple (1932) (as cited in ref. 21) observed an oxidised fat syndrome in dogs kept on diets containing rancid fat. The manifestation of this disorder were described as loss of hair, skin lesions, anorexia, emaciation and intestinal haemorrhages, McKenzie and McCollum (cited in ref. 21) showed that cod-liver oils given to rabbits caused muscular dystropy. They went on to show that vitamin E deficiency resulted in similar symptoms. The oxidised fats in the cod-liver oil were destroying the vitamin *in vitro* and *in vivo*. Hass (1938) (cited in ref. 21) showed that fatty acid peroxides lead to the formation of yellow fluorescent pigments such as lipofuchsin.

In summary, the symptoms of rancid fat toxicity are:

1. *Diarrhoea*. This is probably caused by the formation of fatty acids with hydroxy groups, such as ricinoleic acid, from fatty acid hydroperoxides. Ricinoleic acid is a major fatty acid found in castor oil, whose emetic properties are well known.
2. *Poor rate of growth*. In part this is the result of the diarrhoea. In several of the studies the poor rate of growth was probably a reflection of the animal's distaste for the diet. It is very difficult to persuade experimental animals to eat rancid fat.
3. *Myopathy*. Fatty acid infiltration of the muscle and the replacement of healthy muscle with scar tissue (fibrosis) have been reported in many studies. This usually involves both heart (cardiomyopathy) and skeletal muscle.
4. *Hepatomegaly*. An enlarged liver is a common finding and there may also be hypertrophy of the bile ducts.

5. *Steatitis or yellow fat disease.* The deposition of lipofuchsin pigments in adipose tissue.
6. *Haemolytic anaemia.* The erythrocyte membrane is susceptible to damage by lipid peroxides. Haemolysis of erythrocytes if severe can lead to a mild anaemia, which may explain the normochromic anaemia in Alaskan Eskimos.[22] Platelet turnover may also be increased.
7. *Secondary deficiencies of vitamins A and E.* Peroxidised fatty acids destroy both vitamins A and E in foods. Lipid peroxides also accelerate the rate of turnover of vitamin E in the body, and thus increase the requirement for the vitamin.

Steatitis or yellow fat disease is a naturally occurring disorder in animals living on diets containing oxidised fats, usually of marine origin. The term steatitis is used to describe the inflammatory changes observed in adipose tissue. Adipose tissue and other soft tissues accumulate a yellow/brown pigment called lipofuchsin that results from the reaction of polyunsaturated fatty acid peroxides with ethanolamine phosphoglycerides. It can be induced in experimental animals by feeding fish oils[23] or canned fish.[24] The severity of the steatitis is correlated with the peroxide value of the oil. It can be prevented to some extent by adding extra vitamin E or synthetic antioxidants to the ration.[23]

Dietary linoleic, eicosapentaenoic and docosahexaenoic acids but not α-linolenic acid exacerbate the cardiomyopathy in rats caused by low intakes of selenium or vitamin E.[18] However, cardiomyopathy has been observed in animals fed on low erucic rapeseed oil which is relatively rich in α-linolenate.[25] It is of interest that in toxicity studies of the antioxidant BHT, the incidence of myocarditis in rats was found to decrease with increasing intake of BHT.[26] This implies that relatively low levels of oxidation products lead to myocarditis. There is evidence that rancid fat causes cardiomyopathy in man. This was well documented in the 1920s from work on children fed rancid cod-liver oil.[21]

7.8 Toxicological studies on thermally oxidised fats

A number of lifespan studies have been carried out in rats fed thermally oxidised fat. Oil that had previously been used in cooking, either for long periods or until it was no longer suitable for use, was given as the sole source of dietary fat. Except for a slight decrease in feed efficiency, which was attributed to the poor absorption of the polymers, growth rate, survival and composition and appearance of organs were normal.[27] These studies revealed no major adverse effects from used frying fats, given as 10% of the diet; in these studies the maximum temperature was not higher than 180 °C and the oils were heated for up to 96 h.

It has been claimed that thermally oxidised vegetable oils when fed to experimental animals lead to hypertrophy of the bile ducts and that this is related to pharmacologically active material present in the non-glyceride fraction.[28]

Billek[29] showed that toxicity was related to the polar material in oxidised oil. Sunflower oil used for the production of fish fingers was collected from the factory after normal use and fractionated into polar and non-polar material. Rats were fed either the unused oil, the used oil, the polar fraction or the non-polar fraction at 20% of their diet for 18 months. The polar fraction led to a reduction in growth and increases in the activity of enzymes that indicate liver damage; the liver and kidney weights were also increased. The survival and histology of the tissues were normal. No adverse effects were noted in the animals fed the untreated oil, the non-polar fraction or the used oil.

On the basis of these studies Billek has argued that, provided the polar material does not exceed 27–30% of the total oil, no toxic effects are likely to be observed. It may not, however, be justifiable to extrapolate these conclusions obtained with rats to man.

7.9 Possible long-term effects associated with the consumption of oxidised fats

Epidemiological studies suggest a relationship between the use of both fried and smoked food and cancer incidence.[30] The use of smoked food is associated with an increased risk of stomach cancer, whereas a high intake of fried food is associated with an increased risk of cancer of the breast, prostate and colon. It is to be noted, however, that countries with a high incidence of cancer of the breast, prostate and colon have low rates for stomach cancer and vice versa. Oxidised fats may be involved in both the initiation and promotion of tumour growth.

Fatty acid hydroperoxides are only weakly mutagenic.[31] Malonaldehyde, which results from lipid peroxidation, is both mutagenic and carcinogenic.[32] Polycyclic aromatic hydrocarbons which can result from the pyrolysis of fats are potent carcinogens. The consumption of charbroiled meats by man has been shown to induce drug-metabolising enzymes in the liver.[33] This provides evidence that the amounts present in food are sufficiently great to exert physiological effects. It is virtually impossible to predict what proportions of cancers might be caused by these compounds.

More than 40 years ago Roffo (as cited in ref. 13) heated and oxidised various fats and oils at 250–300 °C for 30 min and then fed them in a bread and milk diet to 1000 rats. He claimed that this resulted in carcinoma of the stomach. Billek[13] claims that Roffo's results have never been confirmed and states that his results have been criticised on the grounds that true malignancy was never shown. Morris et al.[34] also reported gastric adenomas in old rats fed rancid fat. Epoxides from soyabean oil and butyl-9,10-epoxystearate produced subcutaneous sarcomas and pulmonary adenomas in mice.[35] Vysheslavova[36] found a slightly higher yield of tumours in rats fed heated sunflower oil compared with unheated oils. Other studies have failed to show such an effect.[27,29]

Lipid peroxide concentrations increase with age[2] and it has been argued that an inability to control free radical reactions is associated with premature ageing. Evidence supporting this idea includes the observation that antioxidants prolong life in rats[26] and that excessive exposure of skin to sunlight leads to free radical formation and prematurely ages the skin. A shocking example of the effect of free radicals on ageing is the toxic oil syndrome,[37] associated with the 'Spanish oil scandal'. Although the exact cause of this syndrome is unknown, many of the symptoms are characteristic of damage caused by free radicals. Rapeseed oil for industrial use, which was adulterated with aniline, was imported from France into Spain, illicitly reprocessed and sold as edible oil. More than 20 000 people were affected and 340 died; the symptoms were severe respiratory problems, damage to the vascular endothelium and neuropathy.

Atherosclerosis is a degenerative disease affecting the lining of the arterial wall, leading to heart attacks and strokes in the elderly. Lipid peroxides may be involved in atherogenesis. They have been shown to inhibit the production of prostacyclin,[38] a prostaglandin that protects the endothelial lining of blood vessels. Lipid peroxides also increase the tendency of blood to clot by stimulating thrombin generation.[39]

Risk of atherosclerosis is related to high plasma concentrations of low density lipoprotein (LDL) and the cholesterol carried by it. LDL appears to be susceptible to free radical-mediated oxidation both *in vitro* and *in vivo*.[40] Oxidation of LDL increases its uptake by macrophages via a scavenger receptor pathway and renders LDL cytotoxic to cells in tissue culture. Uptake of oxidised LDL by the scavenger receptor pathway may be important in the initiation of atherosclerosis. Malonaldehyde, which is produced enzymically by platelets as well as by autoxidation, also leads to modification of LDL.[2] Modified LDL is not recognised by receptors that take up LDL and so must be metabolised by the receptor-independent pathways. Malonaldehyde-modified LDL also leads to proliferation of smooth muscle cells in the blood vessel wall.

The oxidation products of cholesterol have also been shown to be atherogenic[41] in experimental animals and to cause smooth muscle cells to proliferate. Oxidised cholesterol may either be carcinogenic or promote tumour growth.[3] It has recently been suggested that the high incidence of coronary heart disease amongst Asian men of Indian descent in Britain may be related to their intake of oxidised cholesterol in ghee.[42]

7.10 Conclusion

Oxidised fats are undesirable components of human diets. It is important, however, to realise that the rates of formation of lipid peroxides in the body can be greater than provided in the diet. The most important potential sources of oxidised fat in the diet are most oily fish and high-fat foods which are cooked at high temperatures or exposed to ionising irradiation.

References

1. Logani, M. K. and Davies, R. E. (1980). Lipid oxidation: biologic effects and antioxidants—a review. *Lipids*, **15**, 485–95.
2. Addis, P. B. (1986). Occurrence of lipid oxidation products in foods. *Food Chem. Toxic.*, **24**, 1021–30.
3. Alexander, J. C. (1986). Heated and oxidized fats. In *Dietary Fat and Cancer* (eds C. Ip, D. F. Burt, A. E. Rogers and C. Mettlin), Alan R. Liss., New York, pp. 185–209.
4. Warso, M. A. and Lands, W. E. M. (1983). Lipid peroxidation in relation to prostacyclin and thromboxane physiology and pathophysiology. *Br. Med. Bull.*, **39**, 277–80.
5. Bhuiyan, A. K. M. A., Ackman, R. G. and Lall, S. P. (1986). Effects of smoking on protein quality of Atlantic mackerel. *J. Food Proc. Preserv.*, **10**, 115–26.
6. Rhee, K. S. and Stubb, A. (1978). Oxidative deterioration in vegetable oils: health-food oils versus conventional oils. *J. Food Protein*, **41**, 443.
7. FAO (1962). *Fish in Nutrition* (eds E. Heen and R. Kreuzer), Fishing News (Books), London.
8. Choo, S. Y., Miyashita, K., Miyazawa, T., Fujimoto, K. and Kaneda, T. (1987). Autoxidation of ethyl eicosapentaenoate and docosahexaenoate. *J. Am. Oil Chem. Soc.*, **64**, 876–9.
9. Bhuiyan, A. K. M., Ratnayake, W. M. N. and Ackman, R. G. (1986). Stability of lipids and polyunsaturated fatty acids during smoking of Atlantic mackerel (*Scomber scombrus L.*). *J. Am. Oil Chem. Soc.*, **63**, 324–8.
10. British Nutrition Foundation (1987). *Trans Fatty Acids*, Report of the British Nutrition Foundation's Task Force, British Nutrition Foundation, London.
11. Siu, G. M. and Draper, H. H. (1978). A survey of the malonaldehyde content of retail meats and fish. *J. Food Sci.*, **43**, 1147–9.
12. Newburg, D. S. and Conlon, J. M. (1980). Malonaldehyde concentrations in foods are affected by cooking conditions. *J. Food Sci.*, **45**, 1681–7.
13. Billek, G. (1985). Heated fats in the diet. In *The Role of Fats in Human Nutrition*, (eds F. B. Padley and J. Podmore), Ellis Horwood, Chichester, pp. 164–72.
14. Frankel, E. N., Smith, L. M., Hamblin, C. L., Creveling, R. K. and Clifford, A. J. (1984). Occurrence of cyclic monomers in frying oils used for fast foods. *J. Am. Oil Chem. Soc.*, **61**, 87–9.
15. Wills, E. D. (1980). Studies of lipid peroxide formation in irradiated synthetic diets and the effects of storage after irradiation. *Int. J. Radiat. Biol.*, **37**, 383–401.
16. Halliwell, B. and Gutteridge, J. M. C. (1986). The importance of free radicals and catalytic metal ions in human disease. *Molec. Aspects Med.*, **8**, 89–193.
17. Wilson, R. L. (1987). Vitamin, selenium, zinc and copper interactions in free radical protection against ill-placed iron. *Proc. Nutr. Soc.*, **46**, 27–34.
18. Jenkins, K. J. and Ewan, L. M. (1967). Promotion of myopathy by polyunsaturated fatty acids in cod liver oil. *Can. J. Biochem.*, **45**, 1873–80.
19. Slater, T. F., Cheesman, K. H., Davies, M. J., Proudfoot, K. and Xin, W. (1987). Free radical mechanisms in relation to tissue injury. *Proc. Nutr. Soc.*, **46**, 1–12.

20. Lemoyne, M., Vam Gossum, A., Kurian, R., Ostro, M., Axler, J. and Jeejeebhoy, K. N. (1987). Breath pentane analysis as an index of lipid peroxidation: a function test of vitamin E status. *Am. J. Clin. Nutr.*, **46**, 267–72.
21. McCollum, E. V. (1956). *A History of Nutrition*, Riverside Press, Cambridge, MA.
22. Scott, E. M., Wright, R. C. and Hanan, B. T. (1955). Anemia in Alaskan Eskimos. *J. Nutr.*, **55**, 137.
23. Oldfield, J. E., Sinnhuber, R. O. and Rasheed, A. A. (1963). Nutritive value of marine oils. II. Effects of *in vivo* antioxidants in feeding menhaden oil to swine. *J. Am. Oil Chem. Soc.*, **48**, 357–9.
24. Munson, T. O., Holzworth, J., Small, F., Witzel, S., Jones, T. C. and Luginbuhl, H. (1958). Steatitis ('yellow fat') in cats fed canned red tuna. *J. Am. Vet. Med. Assn*, **133**, 563–8.
25. Trenholm, H. L., Thompson, B. K. and Kramer, J. K. G. (1979). An evaluation of the relationship of dietary fatty acids to the incidence of myocardial lesions in male rats. *Can. Inst. Food Sci. Technol. J.*, **12**, 189–93.
26. Olsen, P., Meyer, O., Bille, N. and Wurtzen, G. (1986). Carcinogenicity study on butylated hydroxytoluene (BHT) in Wistar rats exposed in utero. *Food Chem. Toxic.* **24**, 1–12.
27. Nolan, G. A., Alexander, J. C. and Artman, N. R. (1967). Long-term rat feeding study with used frying fats, *J. Nutr.*, **93**, 337–48.
28. Kaunitz, H. and Johnson, R. E. (1972). Exacerbation of heart and liver lesions by feeding of various mildly oxidised fats, *Lipids*, **8**, 329–35.
29. Billek, G. (1979). Heated oils—chemistry and nutritional aspects. *Nutr. Metab.*, **24**(Suppl. 1), 200–10.
30. National Research Council (1982). *Diet, Nutrition and Cancer*, National Academy Press, Washington, D. C.
31. MacGregor, J. T., Wilson, R. E., Neff, W. E. and Frankel, E. N. (1985). Mutagenicity tests on lipid oxidation products in *Salmonella typhimurium*: monohydroperoxides and secondary oxidation products of methyl linoleate and methyl linolenate, *Food Chem. Toxic.*, **23**, 1041–7.
32. Shamberger, R. J., Andreone, T. L. and Willis, C. E. (1974). Antioxidants and cancer. IV. Initiating activity of malonaldehyde as a carcinogen, *J. Natl. Cancer Inst.*, **53**, 1771–3.
33. Anderson, K. E., Conney, A. H. and Kappas, A. (1982). Nutritional influences on chemical biotransformations in humans. *Nutr. Rev.* **40**, 161–71.
34. Morris, H. P., Larsen, C. D. and Lippincott, J. W. (1943). Effects of feeding heated lard to rats with histological description of the lesions observed. *J. Natl. Cancer Inst.*, **4**, 285–303.
35. Kotin, P. and Falk, H. L. (1963). Organic peroxides, hydrogen peroxide, epoxides and neoplasia, *Radiat. Res.*, **3**, 193–211.
36. Vysheslavova, M. (1968). Carcinogenic effects of overheated fats. *Vopr. Pitan.*, **27**, 63–8.
37. WHO (1980). *Toxic Oil Syndrome, Mass food poisoning in Spain*, WHO, Copenhagen.
38. Salmon, J. A., Smith, D. R., Flower, R. J., Moncada, S. and Vane, J. R. (1978). Further studies on the enzymatic conversion of prostaglandin endoperoxides into prostacyclin by porcine aorta microsomes. *Biochim. Biophys. Acta*, **523**, 250–2.
39. Barrowcliffe, T. W., Gray, E., Kerry, P. J. and Gutteridge, J. M. C. (1984). Triglyceride-rich lipoproteins are responsible for thrombin generation induced by lipid peroxides. *Thromb. Haem. (Stutt.)*, **52**, 7–10.
40. Hiramatsu, K., Rosen, H., Heinecke, J. W., Wolfbauer, G. and Chait, A. (1987). Superoxide initiates oxidation of low density lipoproteins by human monocytes. *Arteriosclerosis*, **7**, 55–60.
41. Imai, H., Werthessen, N. T., Subramanyam, V., LeQuesne, P. W., Soloway, A. H. and Kanisawa, M. (1980). Angiotoxicity of oxygenated sterols and possible precursors. *Science* **207**, 651–3.
42. Jacobsen, M. S. (1987). Cholesterol oxides in India ghee: possible cause of unexplained high risk of atherosclerosis in Indian immigrant populations. *Lancet*, **2**, 656–8.

8 Rancidity in cereal products
T. GALLIARD

8.1 Introduction

Development of rancidity, with resultant loss of quality and acceptability, can occur in cereal products due to degradation reactions at various points along the chain from the grain as harvested and stored, through the different processing operations and into the final products. For the purpose of this review, rancidity is defined as adverse quality factors, arising directly or indirectly from reactions of endogenous lipids, producing undesirable tastes and odours or unacceptable functional properties.

Cereal grains (wheat, maize, rice, barley, oats, millet, etc.) have represented a major source of the food eaten by man ever since he developed from a hunter-gatherer into a farmer. One reason for the universal importance of cereal crops is that the edible portion, the seed, can readily be harvested and then stored for long periods without excessive loss of nutritional value or wastage by disease or senescence. Cereal grains are less prone to spoilage than crops with high water content (roots, leaves, fruits, etc.). Thus, grain which is healthy and sound at harvest and is handled and stored with care, may be expected to retain its usefulness as a food source over many months or even years. However, disease, damage, poor handling, storage and subsequent processing can all induce changes that may lead to rancidity problems in the raw materials or in products subsequently made from them. In Nature, seeds often have to survive adverse conditions before attaining an appropriate environment for germination and growth. Thus, a number of protective features exist to prevent spoilage of cereal grains. Rancidity will be less likely to develop when the lipid materials (e.g. seed oil) are unable to interact with other reactants (e.g. air) or catalysts (e.g. enzymes) because of natural compartmentation within plant cells, or because of diffusion limitations imposed by low water activity. Rancidity is often due to only minor amounts of lipid degradation products and, therefore, the amount of lipid in a product is far less important than its nature and susceptibility to degradation.

In general, rancidity is a greater problem in cereal products derived from wholegrain or those containing bran and/or germ components because, as shown below, these tissues contain the enzymes involved in enzyme-catalysed lipid degradation, and the germ in particular is rich in unsaturated lipids. The dramatic growth in popularity of wholegrain/high fibre products has heightened the importance of rancidity and its control in cereal-based foods.

As with some other types of food, rancidity in cereal products may be due to hydrolytic or oxidative degradation reactions, and often, both. As will be shown below (see also Figure 8.1) hydrolysis may predispose products to subsequent oxidative rancidity.

8.2 Hydrolytic rancidity

In some foods, and particularly in dairy products, hydrolytic rancidity is manifested as a 'soapy' off-flavour due to short-chain (C_6–C_{10}) fatty acids released from the acyl ester forms in which they occur in lipids; such reactions are catalysed by lipase enzymes. The lipids in cereal grains contain predominantly longer fatty acyl chains (C_{16}–C_{18}) and the corresponding free, i.e. unesterified, fatty acids (FFA), have much higher flavour thresholds than the shorter chain acids and, thus, contribute less *per se* to off-flavours. However, hydrolytic degradation of lipids in cereals is important for two reasons:

(a) Polyunsaturated FFA (the predominant fatty acids in most cereal

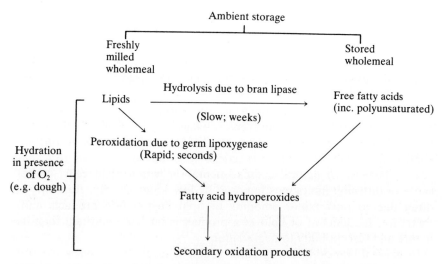

Figure 8.1 Relationship between hydrolytic and oxidative rancidity in wholemeal flour. (Analogous reactions occur in unstabilised milling products from other cereal grains.)

products) formed by hydrolysis of lipids are the precursors of both volatile and non-volatile off-flavours, generated in subsequent oxidation reactions. Enzyme (lipoxygenase)-initiated oxidative degradation requires O_2 and acts predominantly on unesterified polyunsaturated fatty acids, and FFA are commonly oxidised more readily in non-enzymic reactions than are the lipid-bound, acyl ester forms.
(b) FFA are detrimental to the functional properties of many cereal-based products. Hence a maximum FFA content is commonly specified. For example high FFA levels in rice or maize oil causes losses and problems in refining, and the baking quality of wheat flour is reduced when FFA levels are elevated.

Because hydrolytic rancidity in cereal products is enzyme-catalysed, it is important to note some of the unusual properties of the lipase enzymes that are responsible and which exhibit some important differences from other enzymes.

Since the first edition of this volume was published, a considerable amount of research has been carried out in the author's laboratory on the deterioration during storage of wheat milling products. (Details of this work are to be found in a series of publications listed under ref. 1.) This deterioration is initiated by a lipase enzyme. The following text illustrates some important features of this enzyme from wheat, but many of the properties are common to lipases from other cereals.

8.2.1 Wheat (bran) lipase

8.2.1.1 Lipase activity is concentrated in the bran. The lipase that causes hydrolytic rancidity in milling products of sound, ungerminated wheat is found almost exclusively in the bran component. There is an enzyme in wheat germ that has been described as a lipase, and is marketed as 'wheat germ lipase'; however, this enzyme does not hydrolyse the (long chain) triacylglycerols present in wheat and, therefore, is not involved in hydrolytic rancidity in milling products. A true lipase does appear in wheat germ during the early stages of germination, but is not active in products from sound ungerminated wheat.

The difference in lipolytic activity of bran and germ is illustrated in Table 8.1 which shows the accumulation of FFA in bran, germ and mixtures thereof during storage at 20 °C. It can be seen that the FFA level in stored bran is much higher than that in wheat germ. In fact, pure germ (embryo + scutellum) is relatively stable. The fact that commercial wheat germ is unstable and accumulates FFA rapidly is because germ, as produced commercially, contains substantial amounts of bran (up to 50%) and, as shown in Table 8.1, a 50/50 mixture of bran and germ is

Table 8.1 Development of rancidity[a] in finely-milled wheat germ, bran and a 50/50 mixture

Storage period (days at 25 °C)	O_2 uptake capacity (μmol O_2/10 min g^{-1})		
	Germ	Bran	50% Germ, 50% bran
0	2	2	2
21	4	16	27

[a]Rancidity measured as O_2 uptake capacity of aqueous suspensions (shown to correlate closely also with FFA content).[1]

even more unstable than bran alone. This is explained on the grounds that the germ contributes additional substrate (germ oil) for the bran enzyme.

8.2.1.2 Bran lipase is active at low moisture levels. Unlike most enzymes, lipases are active at moisture levels below 5% and well below those present in normal milling products (10–15%). Wheat bran lipase has maximal activity at $a_w = 0.85$ (17% moisture) (Figure 8.2). Thus, FFA accumulate in unstabilised milling products in which other enzymes (e.g. those catalysing FFA oxidation, see later) and microorganisms are inactive, and only operate when more water is added.

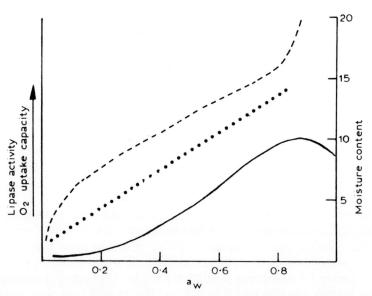

Figure 8.2 Effect of water activity on activity of wheat bran lipase (———) (adapted from ref. 2) and O_2 uptake of wholemeal flour (· · · · ·) (adapted from ref. 1). The adsorption isotherm for wheat flour is shown by the broken line (– – –).

8.2.1.3 Bran lipase is heat stable at normal moisture levels. At normal moisture contents of 10–12%, wheat bran can be held at 80 °C for several days without loss of lipase activity. However, after adding water and heating or by autoclaving, the lipase activity can be rapidly eliminated. In a 50/50 w/w bran–water mixture, lipase activity is inactivated within 10 min at 100 °C; autoclaving (15 psi, 10 min) also inactivates the lipase.

The heat stability and low moisture requirement of bran lipase has led to the development of a simple assay system for the enzyme and this can be used to indicate potential deterioration rates of unstabilised bran-containing products (see Appendix 1).

8.2.1.4 Lipolysis rates increase as particle size is reduced. In bran, brown flours, etc., the rate of FFA accumulation depends upon the bran particle size. This is clearly illustrated in Figures 8.3 and 8.4 which show that, for bran milled to maximise different particle size fractions and then stored at 20 °C, the level of FFA present was highest in the finely milled (< 0.5 mm) material and lowest in the large (> 2 mm) bran particles. We account for the particle size effect on the following grounds: in dry

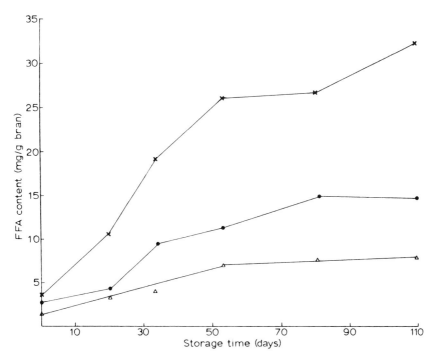

Figure 8.3 Effect of particle size on increase of FFA in wheat bran stored at 20 °C, 65% RH; commercial 'bakers' bran as received (●), ground to pass a 0.5 mm mesh sieve (×), residue on a 2 mm mesh sieve (△).

materials, lipolysis depends upon substrate (oil) diffusing through the material to the bran-bound lipase; the greater the surface area and contact between bran and oil, the greater the reaction rate.

8.2.1.5 FFA levels in flour milling products increase linearly with storage time. Under given storage conditions, there is a steady increase in FFA content of bran (see Figure 8.3) and of wholemeal flour up to a maximum level that depends upon availability of substrate (oil) to the bran lipase. In finely milled (< 0.5 mm) bran, the maximum is reached when at least 90% of the original total triacylglycerol content is hydrolysed; with increased particle size the rate and extent of lipolysis is lower (Figure 8.3).

8.2.2 Lipase activity and hydrolytic rancidity in other cereal grains and products

The lipase activity in some other cereal grains is higher than that in wheat and, again, the bran contains the highest concentrations (Table 8.2). The levels in oats and brown rice are such that the milled products deteriorate very rapidly unless stabilised by a heat treatment. Heat stabilisation is common practice with these materials but is not appropriate for brown and wholemeal wheat flours for baking because the essential gluten protein functionality is destroyed by heating.

8.2.2.1 Oats. Because of the relatively high lipid content and lipase activity of oats, it is standard practice to cook oats during the milling process to inactivate the lipase. A few minutes at 90–100 °C and moisture content > 12% is sufficient to stabilise the products. Without this treat-

Table 8.2 Lipase activities in various cereal grains and milling products[a]

Material	Lipase activity[b]
Wheat, wholegrain	2–4.5
Wheat bran	7
Wheat patent white flour	1–1.25
Oat groat	20
Millet, wholegrain	6–10
Sorghum, wholegrain	6
Rice, brown	11–13
Rice, white (milled)	1.25
Rice bran	20–30

[a]Unpublished data reported by R. Drapron, 7th World Cereal and Bread Congress, Prague (1983). [b]Expressed as oleic acid liberated from a mixture containing ground sample (2 g) and olive oil (100 mg) at 30 °C for 72 h ($a_w = 0.8$).

ment, FFA reach unacceptable levels within 2–3 days. There are varietal differences in lipase activity of oats; a 21-fold difference between the highest and lowest lipase levels was found in tests on 350 cultivars from one location.[3]

8.2.2.2 Rice. Lipase activity is concentrated in the bran layers of rice grain and lipolytic activity is initiated by milling. Thus brown rice and especially rice bran deteriorate rapidly after milling. Rice bran is potentially a rich source of oil (15–20%) but unless adequate stabilisation is carried out at the time of milling, the oil becomes unfit for food use within a few hours of extraction. Numerous processes for rice bran stabilisation have been described but commercial processes involve heat treatment (90–130 °C), including extrusion cooking, at natural moisture levels or less severe heating (> 80 °C) after water/steam addition. To what extent the lipase in ungerminated rice is endogenous or microbial in origin is unclear; a wide range of lipolytic fungi and bacteria are known to colonise rice in the field and on subsequent storage.

8.3 Oxidative rancidity

In all cereal grains, the lipids contain a substantial proportion of polyunsaturated fatty acids; in wheat, barley and maize, linoleic acid (18:2) and linolenic acid (18:3) together account for about 60% of the total fatty acids (Table 8.3). Thus the potential for oxidative rancidity, i.e. the reaction of unsaturated fatty acids with atmospheric O_2 in the presence of catalytic factors, exists and the fact that lipid oxidation is relatively slow in undamaged grain is due to compartmentation of reactants. When cereal grains are processed, the resulting redistribution of components can give rise to oxidation reactions which, in many cases, are rapid and extensive. These may be enzyme catalysed or autoxidation reactions.

Table 8.3 Fatty acid composition of cereal lipids

Fatty acid	Total fatty acids (%)				
	Wheat	Barley	Oats	Rice	Maize (germ)
Palmitic (16:0)	25	12	10	17	8–10
Oleic (18:1)	12	28	59	48	24–50
Linoleic (18:2)	56	56	31	34	34–60
Linolenic (18:3)	4	3	–	1	–

Source: ref. 4.

8.3.1 Enzymic oxidation (lipoxygenase action)

In general, enzyme-catalysed oxidation of lipids in cereal products can be accounted for by the action of lipoxygenase (LOX), an enzyme (more strictly, a group of iso-enzymes) concentrated in the germ (embryonic axis + scutellum) of cereal grains. LOX occurs in all species of commercial cereals, although some cultivars of durum wheat, used for pasta production, have been bred to contain very low levels (see below). LOX acts on polyunsaturated fatty acids containing a 1,4-*cis*-pentadiene group (e.g. 18:2 and 18:3) and preferentially catalyses oxidation of unesterified fatty acids (i.e. the products of lipase activity). The reaction can be summarised as follows:

$$LH + O_2 + Enz \rightarrow \begin{bmatrix} L-O_2 \\ \backslash \cdot / \\ Enz \end{bmatrix} + H^\cdot \rightarrow LOOH + Enz$$

$$X_{red} \xrightarrow{\text{co-oxidation}} X_{ox}$$

where LH = fatty acid substrate, Enz = lipoxygenase, X_{red} and X_{ox} represent reduced and oxidised forms of a co-substrate.

The co-oxidising capacity of the reaction may be of more significance in many cases than the main oxidation reaction. For a given fatty acid substrate, two or more structural hydroperoxide isomers may be formed, resulting in different secondary oxidation products on subsequent reactions of the hydroperoxides. For example, with linoleic acid:

$$CH_3(CH_2)-CH\underset{c}{=}CH-CH_2-CH\underset{c}{=}CH-(CH_2)_7COOH$$

$$\downarrow \text{lipoxygenase}$$

$$\underset{\underset{t}{}}{CH_3(CH_2)_4-\overset{OOH}{\overset{|}{C}H}-CH=CH-CH\underset{c}{=}CH-(CH_2)_7COOH} \quad \text{or}$$

$$CH_3(CH_2)_4-CH\underset{c}{=}CH-CH\underset{t}{=}CH-\overset{OOH}{\overset{|}{C}H}-(CH_2)_7COOH$$

Because LOX is concentrated in the germ, wholegrain and germ-enriched materials show the effects of LOX activity more than refined materials, such as white flour. As part of a major programme in the author's laboratory to study causes of deterioration in wholemeal flour, the role of wheat germ LOX has been studied intensively. In addition to confirming previous knowledge on enzymic oxidation in wheat and other cereal products, this study provided further information on the role of LOX in wheat milling products that is also applicable to other cereals. Therefore, the following text summarises the relevant information for wheat and then refers to other cereal products.

8.3.1.1 Lipoxygenase-catalysed lipid peroxidation in wheat products. Lipid oxidation is relatively slow in sound wheat grain or its milling products, as stored commercially. However, when wholemeal flour or milling products containing wheat germ and bran are mixed with water, the mixtures take up O_2 very rapidly and this O_2 consumption is due almost entirely to LOX-catalysed oxidation of unesterified polyunsaturated fatty acids (18:2 and 18:3). For wholemeal flour suspensions, the rate of O_2 uptake is directly proportional to the amount of unesterified free fatty acid and, because this increases during storage of wholemeal (see section 8.2.1), O_2 uptake of suspensions increases with flour storage time. In a dough made from stored wholemeal, all of the dissolved O_2 is removed within a few seconds of mixing. The reaction products in doughs or suspensions from stored flour are predominantly hydroperoxide derivatives of 18:2 and 18:3. In suspensions of stored wheat bran, hydroperoxides can account for 1–2% of the dry weight. With freshly-milled materials, less oxidation occurs but the products contain higher proportions of the secondary oxidation products, mono- and tri-hydroxy fatty acids.

This rapid and extensive oxidation process in hydrated products has several implications:

(a) Oxidative rancidity, as measured by conventional tests, such as peroxide value, TBA value etc. (see section 2.4) is much higher in hydrated materials (as commonly consumed or processed) than in the dry raw materials.
(b) Doughs of high-extraction flours may contain insufficient O_2 for the oxidation reactions known to be important in dough development for bread making.
(c) Levels of polyunsaturated fatty acids are reduced (particularly when stored materials are used) and replaced with peroxidation products.
(d) Rancid flavours develop rapidly on hydration; for example, with suspensions of wheat bran a very close correlation was found between off-flavour (as detected by a taste panel) and O_2 uptake ($r^2 = 0.998$) or FFA content ($r^2 = 0.993$); see Figure 8.4.

8.3.1.2 Co-oxidation reactions in wheat products. Soyabean flour (containing LOX) has been used in the baking industry for many years; one of its functions is to remove any yellow (carotenoid) colour from white flour

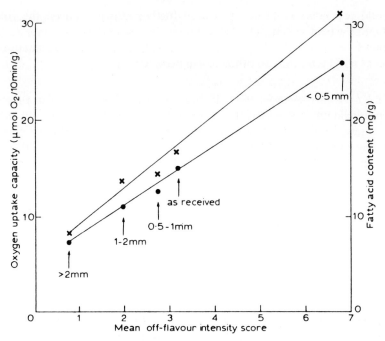

Figure 8.4 Relationship between off-flavour score of bran suspensions (as determined by a taste panel) and either FFA content (×) or O_2 uptake capacity (●). Different particle size fractions (as indicated) of the bran had been stored for 17 weeks at 20 °C, 65% RH.

products. This is an example of the co-oxidation reaction of LOX mentioned in section 8.3.1. High-extraction wheat flours contain sufficient LOX activity (from the germ component) to catalyse bleaching reactions without soya flour addition. For example, wholemeal doughs contain no carotenoid colour. Although not studied in detail in cereal products, it is known that a range of substances, particularly lipophilic compounds, are oxidised by this co-oxidation reaction of LOX. Susceptible compounds include pigments, fat-soluble vitamins (e.g. tocopherols, vitamin A), unsaturated lipids and protein-SH groups. The relevance of this co-oxidation process in cereal products has not yet been fully explored.

8.3.1.3 Measurement of oxidative rancidity in hydrated wheat-milling products. For the reason given above, formation of lipid peroxidation products in unstabilised milling products occurs mainly when these are hydrated (see Figure 8.1). Hence, measurements on hydrated materials are usually more meaningful, since cereal products are usually consumed or processed after hydration.

Measurement of O_2 uptake of suspensions of finely ground milling products is a rapid and simple way of determining the degree of deterioration in the sample; details of the method are presented in Appendix 2. O_2 uptake of extracts in the presence of linoleic acid is a convenient method for direct measurement of LOX activity in aqueous extracts of cereal products; details are given in Appendix 3.

For direct measurement of peroxidation products, methods described elsewhere in this volume are used; a modification of the COP (conjugated oxidation products) method of Fishwick and Swoboda has proved useful in this laboratory.[1] Figure 8.5 shows how the amounts of oxidation products (measured as COP) increase in wholemeal doughs with the storage time of the wholemeal flour.

8.3.1.4 Enzymic oxidation in other cereal products. LOX is present in the germ component of all common cereal grains. In aqueous suspensions of ground grain, a range of secondary oxidation products from LOX-initiated reactions of fatty acids is observed, the proportions of the products varying with the species of grain and, as mentioned above, with the storage time of the ground material. Because lipoxygenase acts primarily on unesterified fatty acids, the extent of enzymic oxidation depends upon the amount of substrate available and this is largely determined by the activity of the endogenous lipase enzyme (see above and Figure 8.1).

On hydration of unstabilised milled oats, a bitter, rancid flavour develops and this has been ascribed to the presence of a trihydroxy fatty

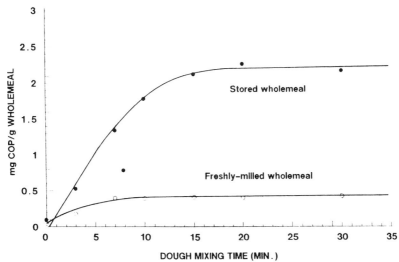

Figure 8.5 Levels of conjugated fatty acid oxidation products (COP) during mixing of doughs from freshly milled wholemeal or from wholemeal flour stored for six months at 20 °C, 65% RH.

acid formed as a secondary oxidation product from LOX-catalysed oxidation of linoleic acid.[5] Other secondary oxidation products are responsible for the 'cardboard' flavour that develops on storage of beer if O_2 is present; LOX from barley is thought to be responsible. It had been assumed that the off-flavour that develops rapidly in ground millet was due to lipid oxidation but recent work has shown that, in this case, deglycosylation of a glycosyl flavone is responsible.

In the manufacture of pasta products from durum wheat, retention of the bright yellow pigmentation from the carotenoids is desirable. Because LOX-catalysed co-oxidation of carotenoids causes bleaching, low LOX levels in the raw material are required; cultivars of durum with very low LOX activity have been bred for this purpose.

8.3.2 Non-enzymic oxidation of cereal lipids

Although enzyme-catalysed oxidation reactions are rapid and extensive under the appropriate conditions, non-enzymic oxidative rancidity does occur in cereal products and can be demonstrated on materials in which LOX has been inactivated.

The initial products of enzymic and non-enzymic oxidation are the same, i.e. fatty acid hydroperoxides. However, there are essential differences between these processes; in particular, non-enzymic oxidation is inhibited by antioxidants that have little effect on LOX. One of the reasons that undamaged grains and cereal germ oils are stable is the relatively high content of tocols, especially in germ and bran. The tocol content in cereal grains ranges from 20 mg kg^{-1} in oats up to over 50 mg kg^{-1} in wheat and maize.[6] Other natural antioxidants are also present, especially in oats.

However, treatments designed to stabilise products with respect to enzyme-induced rancidity may, to some extent, predispose the materials to non-enzymic rancidity. In undamaged grain, much of the lipid is present as oil bodies (spherosomes) and some as components of subcellular membranes. Processing, involving heat and/or shearing will lead to redistribution of lipids, destruction of antioxidants and increased surface area exposed to O_2. For example, steamed oat flakes are relatively stable towards oxidation but on further heating can give rise to unstable products.

Metal ions, probably the main catalysts of non-enzymic oxidation in cereal products, are not only present naturally, especially in bran, but are added deliberately in many cases, for example mineral supplementation of flour, breakfast cereals, animal feed, etc; metal contamination from equipment during processing can also occur.

As oil is confined within discrete bodies in the grain, so also many of the endogenous metal ions are concentrated in organelles, especially in bran and germ which contain around 100 ppm of Fe, a potent oxidation catalyst. Metalloproteins that do not normally catalyse lipid oxidation may do so when denatured by heat/moisture treatment.

Recent studies in Sweden[7] have shown that, during processing, the tocopherol (vitamin E) content of wheat and flour is reduced and, in some cases, almost eliminated. These studies were concerned with nutritional value, rather than rancidity of processed wheat products but the loss of endogenous antioxidant presumably reduced the potential shelf life also. Even without heat treatment, tocopherol levels fall during storage of cereal products; for instance, white flour loses 50% of its tocopherol content on ambient storage for six months.[8]

8.4 Factors affecting rancidity in cereal products

Many of the factors promoting hydrolytic and oxidative rancidity are common and can conveniently be discussed together, particularly since, in many instances, oxidative rancidity is largely a consequence of initial hydrolysis of lipids to generate unesterified fatty acids.

8.4.1 Raw material quality

It is well established that grain which is damaged physically or contaminated with lipolytic micro-organisms will lead to products that have poor stability. Grains subjected to wet pre-harvest weather conditions generally give products that are less stable to rancidity (as well as other problems). For example, fungal contamination of flour can give rise to 'soapy' off-flavour in biscuits due to fungal lipases causing hydrolysis of milk fat in biscuit dough. From evidence mentioned earlier, it is probable that both varietal (e.g. in oats) and environmental factors are involved in the predisposition of cereal products to rancidity.

8.4.2 Processing conditions

It is not feasible to use inhibitors to prevent enzyme-initiated rancidity in foods and, therefore, heat inactivation is the usual approach. However, for reasons given above, the conditions should be sufficient to 'kill' the enzymes, but mild enough to prevent predisposition to non-enzymic oxidative rancidity. In this context, it is worth re-emphasising the fact that sensitivity of enzymes to heat increases markedly with water activity. In some cases, heat treatment would destroy functional properties, e.g. loss of 'vitality' of gluten in dough from heated flour. To circumvent this problem in germ-enriched flour, for instance, wheat germ is cooked separately and then blended with untreated white flour.

8.4.3 Storage conditions

(a) *Temperature.* Apart from the usual kinetic effects of temperature reduction on stability, an additional factor operates in the case of hydrolytic rancidity in cereals. Because lipolysis in dry materials depends upon diffusion of oil in the material, this is restricted below temperatures at which the oil solidifies.

(b) *Water activity.* Unfortunately, lipid degradation occurs at a_w levels well below those pertaining in most cereal products. Because lipases do not require 'free' water for activity, hydrolytic rancidity occurs at moisture levels as low as 5% in flour-based products. Lipoxygenase-catalysed oxidation has been demonstrated at very low a_w levels (0.25) in model systems. However, at least in wheat-based products, the rates at ambient conditions (e.g. $a_w = 0.65$) are several orders of magnitude less than those in hydrated materials. Moisture contents below 5% are usually recommended to minimise non-enzymic oxidative rancidity in cereal-based products.

8.4.4 Atmosphere

Inert gas storage provides no protection against hydrolytic rancidity, although removal of O_2 prevents oxidative rancidity. However, both enzymic and non-enzymic oxidation of lipids require only low levels of O_2 and there are technical and economic problems in achieving and maintaining O_2 levels below 1% in food packaging.

8.4.5 Inhibitors

There are no lipase inhibitors, suitable for use in cereal products, to prevent hydrolytic rancidity. Antioxidants are useful in delaying the onset of non-enzymic oxidative rancidity although they have little effect on lipoxygenase-induced oxidation.

The onset of oxidative rancidity in heat-processed products can be delayed substantially by the use of appropriate antioxidants during the processing. BHA (butylated hydroxyanisole) is particularly effective in products such as roller-dried cereal flakes but legislative and consumer resistance to synthetic antioxidants has led to a search for effective 'natural' antioxidants. Work from the Nestlé laboratories in Switzerland has shown that an extract of rosemary leaves at 300 ppm is effective in delaying rancidity in rice and wheat flakes; vanillin (100 ppm) is a particularly effective antioxidant in roller-dried wheat flakes (J. Löliger, personal communication).

8.4.6 Particle size

The effect of particle size on hydrolytic rancidity was described earlier. Also, oxidative rancidity is enhanced by increasing the surface area both for exposure to oxygen and for coating with redistributed lipids. Obviously, a compromise between stability and acceptable physical and sensory properties is required.

8.4.7 Mixed products

Products containing a mixture of ingredients may deteriorate more rapidly than the individual components; an example of wheat bran and germ mixtures was given earlier. This would occur when one component provided the substrate and another, a catalyst. For instance, oil-rich material and a source of lipase, or material containing free fatty acids and a source of lipoxygenase or a catalyst of non-enzymic oxidation (e.g. Fe, Cu, etc.).

8.5 Conclusions

In some cereal products, stabilisation during processing is essential to prevent rapid onset of rancidity that would render the products unacceptable within a very short time. Examples are milled oats, brown rice and germ- or bran-enriched products. In materials for which heat-treatment would destroy required functional properties (e.g. brown flours for baking), enzyme-catalysed rancidity can be minimised by using low-enzyme activity raw material, avoiding small particle sizes if possible, by attention to storage conditions and by avoiding admixture with oil-rich ingredients.

When heat/moisture treatment is used, this should be sufficient to cause enzyme inactivation and to give the required product characteristics, but it should be noted that such treatments can result in enhanced non-enzymic oxidation and loss of endogenous antioxidants, e.g. tocopherols (vitamin E). The use of added antioxidants may be necessary to give adequate shelf life. Some synthetic antioxidants are effective in dry cereal products as are some plant extracts.

One of the main conclusions from the meeting that prompted the first edition of this book was that careful selection of raw materials and equally careful handling, processing and storage of food materials would minimise problems of rancidity and reduce the requirement for addition of antioxidants in foods. Such a conclusion is certainly valid for cereal grains and their products.

There is much scope for improvement in production and selection of raw materials for end-use, especially with respect to enzyme levels. Also, the fact that most cereal grains are rich sources of 'natural' antioxidants has not been fully exploited.

References

1. Galliard, T. (1986). *J. Cereal Sci.*, **4**, 33–50; 179–82. Tait, S. P. C. and Galliard, T. (1988). *J. Cereal Sci.*, **8**, 55–67; 125–37; Galliard, T. and Collins, A. D. C. (1988). *J. Cereal Sci.*, **8**, 139–46; Galliard, T. and Gallagher, D. M. (1988). *J. Cereal Sci.*, **8**, 147–54.
2. Drapon, R. (1972). *Ann. Technol. Agric.*, **21**, 487–99.
3. Freh, K. J. and Hammond, E. G. (1975). *J. Am. Oil Chem. Soc.*, **52**, 358–62.
4. Kent, N. L. (1966). *Technology of Cereals*, Pergamon Press, Oxford, p. 50.
5. Biermann, U., Wittmann, A. and Grosch, W. (1980). *Fette Seifen Anstrichm.*, **82**, 236–40.
6. Barnes, P. J. (1983). In *Lipids in Cereal Technology*, (ed. P. J. Barnes), Academic Press, London, pp. 33–55.
7. Hakansson, B., Jagerstad, M., Oste, R., Akesson, B. and Jonsson, L. (1987). *J. Cereal Sci.*, **6**, 269–82.
8. Frazer, A. C. and Lines, J. G. (1967). *J. Sci. Food Agric.*, **18**, 203–7.
9. Shipe, W. F., Senyk, G. F. and Fountain, K. B. (1980). *J. Dairy Sci.*, **63**, 193–8.
10. Galliard, T. (1986). *J. Cereal Sci.*, **4**, 179–92.

Further reading

Gardner, H. W. (1980). Lipases, lipoxygenases and hydroperoxides. In *Autoxidation in Foods and Biological Systems*, (eds M. G. Simic and M. Karel), Plenum Press, New York, pp. 447–504.
Chan, H. W. (ed.) (1987). *Autoxidation of Unsaturated Lipids*, Academic Press, London.
Barnes, P. J. (ed.) (1983). *Lipids in Cereal Technology*, Academic Press, London.
Galliard, T. (1987). Lipolytic and lipid oxidation reactions in stored wholemeal flour. In *Cereals in a European Context* (ed. I. D. Morton), Ellis Horwood, Chichester, pp. 462–79.

Appendix 1: Measurement of wheat bran lipase activity in flour milling products

This simple method is based on the observations (a) that wheat bran lipase is active in milling fractions in the absence of added water; (b) in dry materials the enzyme is not inactivated at temperatures up to about 80 °C for several hours; (c) the reaction rate increases with temperatures up to at least 75 °C.

Method

The test material is fine-milled to pass a 0.5 mm screen. A sample (1 g) is weighed into a screw-cap bottle (2.5 cm × 8.5 cm). A solution (1 ml) of high purity (low in free fatty acid) olive oil in hexane (50 mg ml^{-1}) is added and mixed well with the sample. The hexane is then evaporated with a stream of N_2 (until no hexane can be detected; approx. 15 min). To obtain a zero-time value (FFA_0), the fatty acids, together with other lipids are extracted immediately by vigorous stirring (magnetic 'flea') with chloroform–methanol (9:1, v/v; 20 ml) for 20 min. The mixture is then filtered and the FA content of the filtrate is determined by a standard colorimetric (copper soap) method.[9]

To measure the fatty acid produced by lipase activity, another sample of test material is placed in an oven at $75 \pm 1\,°C$ for 5 h, before cooling and extraction with chloroform–methanol and measurement of the fatty acid content (FFA_5) as above.

All determinations (FFA_0 and FFA_5) are carried out at least in duplicate and the results expressed as mg FFA/g test material (natural moisture basis). A solution of pure linoleic acid in chloroform–methanol (9:1, v/v) is used for calibration.

Lipase activity is expressed as the difference (ΔFFA) between mean FFA_5 and FFA_0 values. In the author's laboratory, lipase (ΔFFA) values for wholemeal flour have ranged from 0.3 to 3.4 mg FFA/5 h g^{-1} and for wheat bran from 1 to 10 mg FFA/5 h g^{-1}. The standard error of the difference typically is ± 0.06 mg FFA/5 hg^{-1}.

The lipase assay is valuable in that it can be used on freshly-milled materials to predict the relative deterioration rates on subsequent storage (see Appendix 2).

Appendix 2: Determination of degree of deterioration of flour milling products

It has been shown that a simple measurement of O_2 uptake of aqueous suspensions of bran or wholemeal flour correlates very closely ($r = 0.96$; $p < 0.001$) with measurements of free fatty acid contents.[10] The O_2 uptake values increase linearly with the storage time at ambient temperature and the rate of increase is dependent upon the lipase activity of the bran component. Thus, the method is valuable in checking the state of storage deterioration of milling fractions and is complementary to the lipase assay (Appendix 1) which can predict the subsequent relative deterioration rates of freshly milled materials. As with the lipase assay, the O_2 uptake is carried out on materials requiring no treatment other than grinding.

Method

The test material is finely milled to pass a 0.5 mm screen.

To the reaction cell of an O_2 electrode system (Rank Bros., Cambridge, or equivalent) at 25 °C is added a solution (4 ml) of air-saturated

0.1 M Na-phosphate buffer, pH 6.0. The cell is closed and, after the system has equilibrated to give a steady response, a sample (25–200 mg) of finely milled test material is added and the cell closed again rapidly. The response of the electrode is monitored. It is convenient to use amounts of test material that will produce a depletion of dissolved O_2 of between 25 and 75% in 10 min. The upper (100%) and lower (0%) limits of dissolved oxygen response are set using glucose + glucose oxidase and assumed to represent a range of 0.24 μmol O_2/ml of air-saturated buffer at 25 °C (0.96 μmol in the 4 ml reaction system). The amount of O_2 consumed by the sample in 10 min is calculated accordingly and expressed as μmol O_2/10 min/g material. Values for wholemeal flour range from < 0.5 (freshly-milled) to > 4 μmol O_2/10 min g^{-1} for stored material. Corresponding values for wheat bran range from < 1 to > 40 μmol O_2/ 10 min g^{-1}.

A simplified version of the above method has been developed in the author's laboratory for use in laboratories without research O_2 electrode facilities. For this, a stoppered conical flask containing buffer solution (minimal head-space) and a magnetic stirrer bar is used. Using a dissolved O_2 meter probe, readings of O_2 concentration are taken before and 10 min after addition of test material. Calculations are as above, with appropriate corrections for volume of buffer.

Appendix 3: Assay for lipoxygenase activity in cereal extracts

The following method is used in the author's laboratory, mainly for wheat and its derivatives; the method is also applicable to other materials.

Enzyme extract

Grain, semolina, flour, etc. (1–5 g) is homogenised with ice-cold, 0.1 M imidazole buffer, pH 6.9 (10 ml). The mixture is centrifuged (0–4 °C) at 2000 g for 10 min; the supernatant is then decanted and held in an ice bath.

Substrate

(a) *Stock solution.* Linoleic acid (1 g, 99% pure; Sigma Chemical Co., London) is dissolved in 35 ml of light petroleum (b.pt. 100–120 °C) containing butylated hydroxytoluene (BHT; 5 mg/100 ml) as antioxidant. This stock solution, which contains 0.1 M linoleic acid, may be stored for 3–4 weeks in the absence of light.

(b) *Working solution.* A portion (1 ml) of the stock solution is transferred to a 25 ml beaker and the solvent removed by warming under a current of N_2. To the residue is added 2 M NH_4OH (1–2 ml) and the mixture is heated (60–70 °C) to dissolve the linoleic acid as its ammonium salt. Excess NH_3 is then removed with a stream of N_2 and the solution diluted to 20 ml with 0.1 M imidazole buffer, pH 6.9. This working solution must be freshly prepared each day.

Oxygen electrode

To the oxygen electrode cell (Rank Bros. Ltd., Bottisham, Cambridge), maintained at 25 °C, are added, in the following sequence: 0.1 M imidazole buffer, pH 6.9 (2.0 ml); linoleic acid working solution (0.5 ml). The cell closure is lowered to expel the headspace air and the solution then stirred magnetically. After a steady electrode response is obtained (chart recorder), a sample of the enzyme extract (20–200 µl) is added by syringe and the steady initial rate of O_2 consumption is recorded. Enzyme activity is expressed in units (µmol O_2 consumed/min) and usually related to original weight of sample. The electrode output is calibrated by measuring the response for air-saturated buffer and for oxygen-free buffer (obtained with glucose and glucose oxidase) and assuming that the difference represents 0.6 µmol of O_2 in a 2.5 ml cell. (The oxygen concentration of air-saturated, dilute aqueous solutions at 25 °C is 240 µmol.)

In order to check the proportionality of enzyme activity with the amount of extract added, different aliquots of the enzyme extract are used to give a linear activity–extract volume relationship. Initial experiments with unfamiliar materials should include suitable controls, including assays without added substrate and with oleic acid (not a lipoxygenase substrate) replacing linoleic acid.

Typical values for lipoxygenase activity in wheat range from 0 to 1 unit g^{-1} for white flour or semolina and 2–3 units g^{-1} for grain and wholemeal flour to 20–30 units g^{-1} for commercial wheat germ and higher values for dissected germ.

9 Prevention of rancidity in confectionery and biscuits—a Hazard Analysis Critical Control Point (HACCP) approach
A. FRAMPTON

9.1 Introduction

Over the years UK oil refiners had often been sceptical of the policies and procedures operating in some confectionery and biscuit factories on the handling and use of oils and fats. Whilst now considered history, a quotation from K. Williams' revision of Bolton and Revis 1950, 'Oils, Fats and Fatty Foods' confirms that scepticism.

> 'The question of rancidity probably exercises the mind of the food analyst more constantly than almost any other, either in oils and fats themselves, or in products of which they form a constituent. It is common for a manufacturer to accuse the oil which is used in preparing such products as being the primary cause, and to hold the manufacturer of the oil to blame for want of care in refining of his product. Such an accusation can scarcely ever be substantiated conclusively, as, in admixture, the oil is at the mercy of any lipase producing organisms which may have happened to gain access to the final product during manufacture or of adventitious moulds or other harmful influences. However pure or highly refined an oil may be, such attributes provide no protection and the cause of rancidity in edible fatty materials such as chocolate, biscuits and confectionery will be found with far greater likelihood to lie in faulty methods of manufacture of the product concerned, and more particularly in the presence of an excessive percentage of moisture coupled with unsound constituents'.

Just before the demolition of the William Crawford biscuit Factory at Liverpool in 1986, copies of a letter dated 1931 and documents from Schaal to the Strietmann Biscuit Co., Cincinnati, now part of the Keebler Co., USA, were found in the building. The letter related to problems of rancidity, in particular, in coconut oil-based products. It was of particular interest as the test now known as the Schaal Test[1] was described. It appeared that rancidity was then a serious problem. Reference is made to part of the documents attached to the letter issued by the Biscuit and Cracker Manufacturers' Association.

> 'Soapy tastes in biscuit products arise from time to time in almost every biscuit plant. Many complaints on soapy goods have come to the attention of the

Bureau. In every instance the soapy taste had developed with sandwich or certain sweet good pieces in which coconut fat had been used. Coconut fat stands out among all the fats as the one most prone to become soapy'.

Fifty years on, rancidity is not a problem in the UK confectionery and biscuit industry. What have we learnt from the past? Can the developments and the experiences of the past help in the hazard analysis critical control point (HACCP) approach?

The author considers it extremely helpful, and in some instances essential, to be aware of past developments which have reduced or even eliminated the risk of specific hazards being a threat to product safety and/or acceptability.

This awareness can now be readily achieved from the vast sources of technical information available from data banks etc. The experience of employees in the industry should also be utilised.

How has this sector of the food industry achieved such a position? A bird's eye view of progress over the last thirty years in a number of important areas is now considered.

9.2 Ingredients

9.2.1 Oils and fats

Major improvements in the oxidative stability of refined oils and fats have been achieved as a result of a number of improved processing techniques. Three in particular are identified.

9.2.1.1 Hydrogenation. Catalysts for selective hydrogenation with the ability to maintain levels of $C_{18:3}$ well below 1.0% (with a target of 0.5%) whilst still maintaining functional properties have been developed. Expertise in the hydrogenation of marine oils has enabled them to be used as a major constituent of dough fat blends. The reversion flavour problems of the 1960s are not now experienced.

9.2.1.2 Use of sequestrant. In the early 1970s citric acid addition was introduced. This significantly improved the oxidative stability by removal of trace transition metals by chelation. The user now expects to be supplied with refined oils and fats with levels of:

- $Cu \ll 0.05$ mg kg^{-1}
- $Fe \ll 1.00$ mg kg^{-1}
- $Ni \ll 0.80$ mg kg^{-1}

In the UK citric acid may be used as a processing aid and is not required to be declared (see 1980 UK Miscellaneous Additives Regulations/1984 UK Food Labelling Regulations). Normal usage rate is at 50 mg kg^{-1}.

9.2.1.3 Activated bleaching earths. The understanding of the role of bleaching earths in the production of quality oils and fats has resulted in the significant development of earths which exhibit specific performance characteristics dependent on pore structure, particle size and surface area.

As refineries developed, much progress was made in the storage and heating facilities for oils and fats. The use of stainless steel has contributed its obvious benefits, as has the phasing out of direct and trace heating in preference to oil or water jackets. Although some new refineries are using hot water coils these changes have eliminated the risk of 'hotspots'.

The use of nitrogen sparging and blanketing has extended shelf life, and considerable improvements have been made in deodorisation.

All these changes have contributed to the supply of excellent quality refined oils and fats to the food industry.

9.2.2 Stabilised cereals (e.g. oats)

The introduction of steam treatment (stabilisation) in oat processing to inactivate lipase has been recognised by the biscuit manufacturer as the most important development in oat technology in the reduction of incidents of rancidity. The stabilisation conditions are normally 2–3 min at 96–100 °C.

Prior to the early 1950s,[2] when it was first introduced in the UK, the effectiveness of kiln drying (in which the moisture content of 14–20% was reduced to 4–8%) was often unreliable for inactivation of the lipase. As in the USA, lauric fats were much in use in dough fat blends and many incidents of soapy rancidity were experienced even until the late 1960s. Oat-based doughs at pHs of 7.4–7.6 and moisture content > 10% provided almost optimum conditions for lipase activity, although the dough temperature of 70–75 °C retarded the rate of reaction.[3]

Controlled storage conditions and selection of lower free fatty acid (FFA) oats have now further reduced the risks (typical FFA values of 3.0% of fat phase). However, the green shelling process (removal of husks immediately after stabilisation) may give rise to the risk of oxidation due to the higher temperatures required for flavour development. Consequently oat products produced by this process may lack the characteristic nutty flavour as they are kilned at a lower temperature to minimise oxidation.

9.2.3 Oleo resin extracts

The availability in the 1950s of standardised oleo resin extracts of herbs and spices on a range of carriers such as dextrose, salt, etc., eliminated the risk of lipase activity from the high levels of moulds, sometimes $> 10^5 \text{ g}^{-1}$ and other micro-organisms present in natural spices and herbs. These extracts are also available as dispersions in vegetable fat or oil. To use natural spices and herbs in lauric-based creams offers considerable risk of rancidity if the water activity is greater than 0.3.[4]

9.3 Packaging materials

Packaging, as is to be expected, plays a vital role in product acceptability. It may or may not, because of marketing considerations, exclude light, but the format is essentially designed to prevent or retard the access of water vapour and oxygen.

From the time that confectionery and biscuits were first packaged the materials used have either presented a risk from atmospheric oxidation or taint. However, the packaging industry in the UK, in particular the biscuit sector, experienced a revolution in the types of packaging that became available in the decade between 1955–1965. From the sale of loose biscuits in greaseproof-lined tins or packaged in paper/wax/tissue laminates to the modern PVC-coated polypropylene package, the biscuit industry progressed to packaging applications with almost no risk of rancidity with respect to packaging. The phasing out of greaseproof or glassine-lined trays in favour of polystyrene or PVC-formed trays considerably reduced the risk of rancidity which often arose from a combination of poor grease resistance, high levels of trace metals in the greaseproof and the type of fat phase. Storage at elevated temperature was often implicated.

The move to one-piece PVdC-coated reverse-printed polypropylene biscuit wrappers in the mid 1970s was instrumental in further reducing the risks of rancidity. Earlier formats had included printed vegetable parchments or greaseproof with an unprinted or printed PVdC-coated cellulose film overwrapper. A good grease resistance and low levels of trace metals in the parchment or greaseproof and printing inks were critical for maintaining low risk. Improved wrapping machines and more precise temperature control has enabled an increased sealing effectiveness to be achieved.

9.4 Equipment

Spurred on by the advice and often constructive criticism of the oil refiner, perhaps after an unfortunate incident most UK chocolate and biscuit manufacturers have made significant advances in the storage and handling of oils and fats. Gone are the square tanks that did not drain in normal use. Gone is the evidence of polymerised oil due to direct electrical localised overheating and lack of temperature control.

With the increased use of epoxy-coated mild steel and stainless steel, dish-shaped vessels with a total drainage facility are now the norm, being used, in particular, for storage of butter oil. The increased use of oil or water jacketed heating systems for vessels and pipelines with precise temperature control has been a major advance.

There has been an increased use of closed implant cleaning (CIP) systems for vessels used for aqueous dough fat blends and butter oil, and in nitrogen blanketing for the storage of the latter. The removal of copper and copper alloys from any functional equipment that would have direct contact with the oil or fat has also been of great benefit.

There has been a diminishing use of grid melters with the move to bulk deliveries; however, some low volume speciality fats continue to be melted into tank storage by this means. Fortunately, these materials are in the main very stable. Grid melters operating at low temperatures (40–60 °C) have come back into fashion for melting down butter from EC stocks in the manufacture of butter oil. The mantle method of heating drums seems to have disappeared. Gone is the 'day of the spade' for drums have been superseded by cartons. However, drums are still being used for liquid oils such as refined deodorised (R.D.) sunflower and cottonseed oils, etc.

9.5 Processing

9.5.1 Commercial buying policy

The transition from purchasing on a spot basis or as parcels to buying to an agreed material buying specification, in particular, on a contract basis, has resulted in materials being supplied more consistently within well defined parameters.

9.5.2 Reduced stock levels

In recent years the use of nominated fat blends and increased flexibility of supply has facilitated the reduction of stock levels, thus ensuring processing generally within 2–3 days of delivery. Whilst the benefits of lower stock levels and good rotation have not been quantified it has not been considered necessary to use antioxidants to minimise oxidation during storage before use in manufacture.

9.5.3 Reduction in blending at receiving factories

Until the advent of Malaysian palm oil as a major constituent of dough fat blends in the late 1960s, it was not uncommon to blend five or six fat components in a formulation. Some of the components would be added at low levels to 'fine tune' the dilatation values (solids profile) and consequently the usage rate would be much slower, thus introducing risks of oxidation during the longer storage period.

The blend would, of course, be required to be homogeneous within a given production cycle, with mixing at the time incorporating some oxygen to the blend's detriment. Some blending is still being carried out but with two or three components at the most.

9.5.4 Increased use of bulk deliveries

Based on commercial justification the resultant move from boxes or drums to bulk deliveries is always advantageous in reducing risks of rancidity and should be phased in as soon as the usage rate permits. However, one should never fall into the trap of using a bulk system to facilitate ease of handling when the usage rate is so low that the turnover of the stored oil or fat well exceeds the specified shelf life.

9.6 Shelf life

9.6.1 Pursuit of freshness — 'own label' policy

Whilst the biscuit industry has for a long time operated systems, some more sophisticated than others, to regulate stock levels of finished products in depots and to monitor quality by Age-of-stock reports, it has been the growth of the own label trade (32.5% of the UK market, 476 693 tonnes in 1987) which has had the major influence on the 'age' of product at the point of sale.

The designated shelf life of own label products is often much less than the branded equivalent. Combined with their own distribution network the products are purchased and consumed in much fresher condition. Some own label stock requirements are to precision order whilst branded products are to national stock levels. These factors, combined with improved packaging, have considerably further reduced the risks of rancidity being detected in finished product within its shelf life.

9.6.2 Minimum durability—'best before'

Coding of production for identification and investigational purposes in the event of a consumer complaint or recall has been carried out in numerous ways for many years, sometimes to confuse the competition. The requirement for open date marking as required by the UK Food Labelling Regulations 1984 (see Chapter 12) has now even further reduced the risk of old stock being on sale. Prior to this regulation being enforced, tins of rancid biscuits which were a few years old would often surface as complaints as there was no sale or return on tins intended for seasonal sale. Progress over the last thirty years has been remarkable and many lessons have been learnt from the mistakes of others; but perhaps there are still many mistakes being made. The confectionery and biscuit industries are now less distinctive than they were some thirty years ago. Both sectors market products which are either pseudo-biscuit or pseudo-confectionery. They now use ingredients that are common to both, and the risks of rancidity are most probably of the same order for each sector.

Changing market conditions, as a result of consumer demands for 'healthy eating' products, are now leading to the requirement for product fat phases to be lower in saturates, with consequential potential reduction in oxidative stability. Other consumer pressures exacerbate this, like the requirement for lower calorie products, perhaps based on higher water activity systems and the increased use of oats and other ingredients with potential lipase activity if incorrectly processed. Perhaps, looking further ahead, there may be a move back to more 'environmentally friendly' materials such as paper-based packaging. We may even see the removal of some antioxidants from the permitted list. All these factors add up to a potential increase in the risk of rancidity.

It is suggested that the Hazard Analysis Critical Control Point (HACCP) approach used so successfully in assuring microbiological safety in recent years is the required philosophy to prevent rancidity in our confectionery and biscuit products. The principles of HACCP[5] are well known and only the relevant definitions need be briefly outlined.

9.7 HACCP

What is HACCP and how can it be applied to the prevention of rancidity? First, some definitions are presented.

(a) *Hazard analysis*: Any system which analyses the significance of a hazard to product acceptability or consumer safety.
(b) *Hazard*: A potential to cause harm to the product or consumer.
(c) *Critical control points*: A location, stage, operation or material

(ingredient or packaging) which, if not controlled, provides a threat to product acceptability or consumer safety.

In any HACCP approach it is necessary for the following to be established:

1. Description of hazard.
2. Identification of all the critical control points.
3. Identification of control criteria and agreement on monitoring frequency.
4. Quantification of level of concern.
5. Agreement on procedures in the event of deviation.
6. Verification procedures.

9.7.1 Product acceptability

The development of rancidity in confectionery and biscuit products is a threat to their acceptability by the consumer. Rancidity harms the organoleptic quality of the finished product by manifestation of off-odours and off-flavours. The threat to the safety of the consumer from the products of oxidation and hydrolysis are considered in chapter 7.

9.7.1.1 Oxidation. Oxidised flavour is an accumulative term describing off-flavours which may have characteristics such as metallic, fishy, cucumber, cardboard, tallowy, mushroom and painty notes, etc. It should be borne in mind that not all oxidised flavours are undesirable, e.g. the presence of creamy notes in butter from *cis*-4-heptenal. Bitterness is associated with enzymic oxidation in cereals.[6]

9.7.1.2 Hydrolysis. Soapy tastes arise from myristic and lauric fatty acids from triglyceride hydrolysis. Cheesy notes may occur in milk fat.

9.7.1.3 Reversion. The early stages of reversion are often characterised by buttery notes, in particular, in soyabean oil. Other reversion flavours have been described as green beany, weedy, melon, hay and fish-like.

9.7.2 Description of the hazards

There are two hazards associated with rancidity, atmospheric oxidation and hydrolysis.

Atmospheric oxidation is a process of oxidation when air comes into contact with an oil or fat, activated by some external factor such as heat, light, moisture, trace metals or lipoxygenase. Hydrolysis is a process of decomposition of an oil or fat into its constituent fatty acids and glycerol.

Whilst oils and fats may be hydrolysed by heat and moisture, most hydrolysis is brought about by enzymic reaction. Soap residues promote hydrolysis of lauric fats in the presence of moisture. Note that the soapy flavours which result from the action of lipase on lauric-containing oils and fats may also be brought about by the presence of sodium bicarbonate used as a raising agent,[7] and materials with high free fatty acid levels (e.g. oatmeal) producing the sodium salts of fatty acids, which are soaps.

9.7.3 Identification of critical control points

A critical control point is any stage in the manufacture of a product to the point of sale where there are factors which activate or influence the rate of oxidation or hydrolysis. The specific risk factors are as follows:

9.7.3.1 Oxidation factors.

Heat – Rate of reaction doubles with approximately 10 °C rise in temperature.
Light – Can increase initial rate of reaction by up to 10 000 times in direct sunlight.[8]
Trace metals – Catalyses the oxidation by enhancing the formation of free radicals.
Moisture – At high levels in a system acts as a pro-oxidant hydrate and solubilizes trace metals.[9]
Enzymes – Cereal and other plant tissue lipoxygenases oxidise polyunsaturated fatty acids such as $C_{18:2}$ and $C_{18:3}$.
Fungal growth on storage – Ketonic or perfume rancidity. Methyl ketones produced by partial oxidation by specific fungi of intermediate chain length fatty acids (C_6–C_{12}) in materials such as butter and lauric-based foodstuffs.[10]
Pseudo-enzymic activity – Proteins containing transition elements which when denatured may make the metal ions available for catalysis.

9.7.3.2 Hydrolysis factors.

Enzymes – Ingredients containing active lipases.
Water activity (a_w) – Lipases are active at a_w as low as 0.3. Acker has reported activity at $a_w = 0.1$.[11]
Alkali/soap – Catalyses reaction with lauric fats at low levels of moisture. Free alkalinity may give rise to soapy flavours (> 50 mg kg^{-1} as sodium laurate).[12]
pH – Optimum and range for activity.
Temperature – Optimum and range for activity.

In any enzymic hydrolysis, the combination of concentration of enzyme and substrate a_w, pH, temperature and the presence of inhibitors will determine the rate of activity. *Note* that there are few specific inhibitors of lipase but none are permitted for food use, so this possibility should be disregarded.

9.7.4 Critical control points

- Formulation
- Ingredient
- Process
- Finished product
- Packaging
- Distribution/point of sale

9.7.4.1 Formulation. There are two major aspects, new product development and cost reduction exercises.

9.7.4.1.1 New product development. Those developing new products should recognise their responsibility to design out rancidity risks by hazard analysis risk assessment (HARA) and risk avoidance. However, some marketing groups are too often prepared to take risks. In a HACCP approach there should be an independent arbiter, i.e. a quality assurance function.

Product development technologists should be made fully aware of rancidity hazards and the risk categorisation of the ingredients they propose to use in the development. This should also extend to the awareness of risks in the process, packaging format and shelf life. Often the cost contribution will influence the choice of the ingredient.

Some typical questions to be considered are:

- Should animal fat or vegetable fat be used?
- What is policy on use of antioxidants?
- Filling fats—should they be lauric or non-lauric?
- Should premium grade nuts or second grade nuts be used?
- Which nuts (type/origin)?
- Should natural spices or oleo–resin extracts be used?
- Should butter fat or butter flavour be used?

9.7.4.1.2 Cost reduction exercises. Any ingredient to be evaluated as an alternative material in an existing formulation should be automatically scrutinised for its risk to product acceptability and consumer safety. If any of the essential criteria cannot be met, its introduction should not be progressed, even if there is a cost justification.

9.7.5 Ingredients

Ingredients may be conveniently categorised by their degree of risk into three main groups.

High risk

- Ingredients that will readily undergo oxidation if subjected to heat, light or catalysis by trace transition metals.
- Ingredients that readily undergo hydrolysis in the presence of active enzymes or alkali.
- Ingredients which contain active enzymes.
- Ingredients which may, in normal seasons, not present a risk but due to exceptional wet weather conditions during harvesting, possess a very high mould count.

Medium risk

- Ingredients which are subjected to processing which deactivates enzymes, such as roasting, boiling, steam treatment, pasteurisation, etc.
- Ingredients which have known history of a detrimental change of state.
- Ingredients which undergo processing in their manufacture which removes the active enzyme.

Low risk

- No evidence of, or reason to suspect.

Figure 9.1 identifies some implications of risk categorisation. An ingredient cannot be correctly categorised without a comprehensive knowledge and understanding of its properties. Regular visits to suppliers are the only effective way to 'get to know' the ingredient. Unfortunately, this may not always be practical. Details such as origin, nature and composition, process, and environmental and seasonal factors should all be embodied in the material buying specification. Some examples of these important aspects of ingredient characteristics are discussed further.

9.7.5.1 Origin. The country of origin of the raw material and of the ingredients produced from the processed raw material is often of significance. Nuts from various origins may differ in their unsaturated fatty acid composition, with subsequent differences in their oxidative stabilities, for

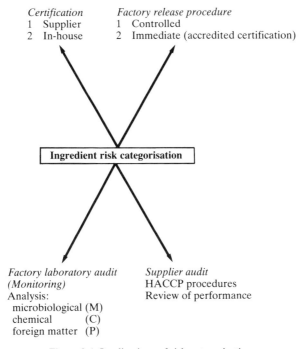

Figure 9.1 Implications of risk categorisation.

example in hazelnuts, their linoleic acid content depends on the origin and variety. Higher levels of linoleic acid reduce the induction period.[13]

9.7.5.2 Nature and composition. It is important that the nature and composition of every ingredient is clearly defined in unambiguous statements, for instance the use of refined soyabean and rape seed oils in dough fat, cream fat and spray formulations should only be as the partially hydrogenated form with levels of $C_{18:3}$ less than 0.5%.

In nuts, the state of maturity can be of some importance. Immature peanuts can contain levels of up to 0.3% linolenic acid, whereas mature peanuts contain approximately 0.02%; this renders the former more susceptible to oxidation.[14]

9.7.5.3 Process. The essential processing steps in the manufacture of the ingredient need to be known. Hydrolytic rancidity problems have been experienced in nougat-type confections produced from 'high boil' syrup, egg albumen-based frappe and lauric fat. In this particular instance the egg albumen had been treated with pancreatic lipase to remove traces of yolk fat, and enzymic activity was carried through into the ingredient as used. To eliminate this risk, it would be necessary to specify no lipase treatment of the egg albumen, or if this was not practical, the avoidance of lauric fats and desiccated coconut in formulations.

9.7.5.4 Environmental and seasonal factors. Cereals, and wheat in particular, harvested during a wet summer are often heavily contaminated with moulds; as a result, the flour milled from them may have a high lipase activity. Such was the experience of the UK biscuit industry in 1957, when a considerable number of soapy rancidity problems were evident.

Halton *et al.*[15] reported that moulds and other fat splitting organisms should be present at less than $10\,000\,g^{-1}$ to avoid the risk of rancidity. Such harvesting conditions may not arise regularly, but in a HACCP approach this risk factor should be identified in the material buying specification for flour.

In the event of further incidents, steps should be taken to reduce the risk by the removal of lauric fats from dough formulations, and to operate a more stringent control on dough standing time. Implications for doughs containing desiccated coconut as a characterising ingredient would also have to be considered.

9.7.6 The process

The author has likened the prevention of rancidity to a fight that never ceases, concluding that the 'fight' is at the beginning.

The technical fight is far too often with the engineers, accountants and marketing departments, in order to establish the essential requirements for plant and product design so as to avoid rancidity. There is more than adequate past experience to emphasise the need for specific materials to be used, making the design of equipment, correct heating facilities, correct storage capacity and cleaning facilities, and so on, important.

To win this fight enlightenment is needed and commitment of senior management to a HACCP approach. The plant and the process should be designed with risk avoidance, product acceptability and safety as priority considerations, including such aspects as storage conditions, stock rotation, mixing and standing times, baking, roasting, intermediate processing such as creaming, etc.

9.7.7 Finished product

The finished product quality criteria such as appearance, texture, flavour profile, dimension and weight will usually be well defined and any defect in these parameters should be evident from organoleptic assessment or physical measurement. Unfortunately, the threat to product acceptability and safety posed by the development of rancidity is not always apparent at the time of production. Therefore, parameters which may influence the activity or rate of reaction must be identified, for example water activity, enzyme activity, trace metals, induction period (Rancimat) and moisture.

9.7.8 Packaging

The changes in packaging resulting in the reduction of incidents of rancidity were referred to earlier. In the use of paper-based materials it is, as with ingredients, important to get to know the materials by visits and regular contact with suppliers. Fortunately the factors which activate or influence the rate of oxidation are few. These are:

(a) traces of transition metals in paper and inks,
(b) grease penetration which permits an increased surface area of oil or fat to be available for oxidation,
(c) exposure to light.

An isolated incident of rancidity occurred some years ago in a low volume but well established semi-sweet biscuit exported to Scandinavia. Samples of the complaint were returned for examination.

The pack was a two-piece combination of printed vegetable parchment and unprinted MXXT/S cellulose film overwrap with end-seals. The packaging format had been in use for some years with no previous incidents of rancidity. The product contained about 16.0% fat which is typical for a semi-sweet formulation. The dough fat formulation at the time was in use for the home trade, and was such that adequate stability could be expected. No evidence of rancidity was detected in the retained shelf-life samples. Returned packs exhibited grease penetration over a wide surface area of the printed vegetable parchment, in particular, in the area of the company logo. Typical oxidative rancidity off-odour was in evidence from the parchment and from the surface of the biscuits. In the course of investigation it was established, much to the consternation of many, that the company logo, which had been printed in a quality up-market image in a gold colour, was found to contain fine bronze powder with a copper content of about 90%. Whilst the precise history of events could not be ascertained, it was most likely that the product had been subjected to very warm conditions which induced fat penetration into the printed parchment and that oxidative rancidity of fat was catalysed by the copper metal in the company logo. The incident illustrates the vigilance required to check every aspect of packaging. Needless to say, the bronze logo was phased out immediately.

Cost reduction exercises or the introduction of window packs may result in reduced ink coverage, thus making the product more vulnerable to photo-oxidation. Blue print has been associated with incidents of rancidity in the past.

9.7.9 Distribution/point of sale

The HACCP approach would not be complete if this critical point were to be ignored. Improvement in distribution and awareness of storage conditions required for products has further reduced the incidence of rancidity.

As described earlier, pursuit-of-freshness and 'best before' have had a tremendous influence on product quality at point of sale. However, some shopkeepers still persist in exhibiting products in window displays.

Depot storage conditions for temperature and humidity should be established. Guidance on recommended storage of products documents should be issued to all retail outlets. Sales personnel can still prevent incidents by advising and educating independent shopkeepers of the risks. In general, the multiples and cash and carry outlets have this under control.

9.7.10 Identification of control criteria and monitoring frequency

In most medium to large confectionery and biscuit companies, it would be expected that the essential control criteria have been identified and monitoring frequencies established. The essential control criteria would be incorporated in material buying specifications and product manufacturing specifications.

The control criteria are those which measure:

1. The state of the ingredients and packaging material. For example, the Rancimat induction period of the oil or fat, or the grease resistance of the cardboard carton.
2. The factors which activate or influence the rate of oxidation or hydrolysis, such as the storage temperature and storage period before use of the oil or fat.

Typical control criteria in the material buying specification for a vegetable oil could be:

- Rancimat induction period
- peroxide value
- totox value
- linolenic acid level
- trace metal content, e.g. Cu, Fe, Ni
- free fatty acid content

Currently, the biscuit and confectionery industries enjoy, in the main, very consistent quality oils and fats with excellent oxidative stability. However, if there is a move to the increased polyunsaturated/saturated fat ratio recommended by the medical experts, oxidative stabilities will be significantly reduced, in some instances by up to 40%. To reduce the risks of rancidity in such circumstances a more integrated approach will be required in correlating ingredient quality, storage conditions, storage time before use and the shelf life of the finished product. Resulting packaging formats such as 'window' packs may not be practical.

Policies and procedures must be developed on a practical basis and implemented with commitment if a HACCP approach is to be meaningful. Monitoring frequency should always be related to the real level of concern and the test schedule and monitoring frequency should be incorporated in the laboratory manual. The monitoring frequency may be relaxed subject to confidence level.

A HACCP approach will inevitably interfere with production schedule, but commitment to risk avoidance is essential. Indeed, no high risk ingredients should be released to production until cleared by the laboratory.

9.7.11 Quantifying level of concern

This is an expression of the seriousness of a failure to control a critical control point derived from the knowledge of oxidation and hydrolysis and the probability of it occurring. Reference was made to the risk categorisation of ingredients which should also be applied to packaging materials, etc.

9.7.12 Agreement on procedures in the event of a deviation

To minimise the risks of rancidity development, policy and procedures for any deviation from the specified criteria for control (tolerances and limits) for all critical points must be established and agreed with all relevant departments. The 'no go' limit (the absolute maximum not to be exceeded) is the most realistic approach, but as rancidity is usually caused by a combination of factors it may be that more information will be required before a judgement is made.

A no go policy is more easily operated on incoming deliveries of materials, e.g. a delivery of oil with the peroxide value exceeding the buying specification value of $1.0 \text{ mequiv kg}^{-1}$ would automatically be rejected before unloading. However, a quantity of fat which has exceeded its agreed shelf life in storage would not necessarily be rejected and returned for reprocessing, but would be referred to the laboratory for flavour assessment, and peroxide value and induction period determination before a decision on its use is made.

9.7.13 *Verification*

In any HACCP system it is essential to be able to verify that the system is functioning efficiently and effectively. Audit procedures, which are preferably independent, on the systems are a necessity.

An integral part of the on-going local audit is organoleptic evaluation of finished product within, and for a limited time beyond its shelf life. The evaluation procedure should be able to detect any organoleptic changes which would affect product acceptability which may have developed over the designated shelf life as a result of the inherent properties of the ingredients, process, finished product, packaging and storage conditions. For effective verification there is need for the development and utilisation of analytical techniques to identify organic compounds produced in foodstuffs at the onset of rancidity. The UK Food Research Association has recently published a confidential report[16] on the use of GC/MS for the detection of such compounds. Another approach may be the use of the Rancimat, to measure the oxidative stability of the finished product and to correlate it with the quality data of the ingredients.

The use of objective measurement has its obvious benefits:

1. Establishment of shelf life for new products,
2. Establishment of shelf life for existing product in which the ingredients or process have been changed.
3. Verification of suspected rancidity,
4. Confirmation that product is completing its shelf life.

9.8 Conclusions

The success of a HACCP approach depends on commitment from the senior management down through all levels of staff. It requires education at all levels, and an awareness of the significance of factors that influence oxidation and hydrolysis and the implications of deviations from the essential criteria of the control points.

The challenge (see section 9.1) the oil refiners gave to the confectionery and biscuit industries has been met, but it has taken a long time to achieve. The mistakes of others have been learnt from. Our understanding of oxidation, hydrolysis and reversion has increased. We know why it happens!

It is more likely that the confectionery and biscuit industries will move into a new era of risk. The cookie industry in the USA, for example, has already seen a move to reduce the level of saturates in fat formulations. It may not be too long before the UK and Europe follow! The HACCP approach, as it evolves, will enable us all to maintain our position of success.

References

1. Joyner and McIntyre (1938). Schaal test. *Oil and Soap*, (Appendix), **15**, 184. Extract from Schaal letter 1931.
2. Kent, N. L. *Technology of Cereals*, Pergamon Press, 3rd edn, p. 168.
3. Matlashewski *et al.* (1982).
4. Minifie B. W. (1970). *Chocolate, Cocoa and Confectionery*, 412.
5. ICMSF (1988). Application of the hazard analysis critical control point system to ensure microbiological safety and quality. *Micro-organisms in Foods*, Blackwell Scientific.
6. Von Bierman *et al.* (1980). Formation of hydroxy acids. *Oat Technology*.
7. Kent, N. L. *Technology of Cereals*, Pergamon Press, 3rd edn, p. 167.
8. BFMRA Lit. Survey No. 4 (1968). *Effect of Light on Food Stuffs*, Lee, 1938.
9. Bailey, Y. Y. *Industrial Oils and Fats*, John Wiley, 4th edn, p. 147.
10. Kinderlerer, J. (1984). *J. Sci. Food Agric.*, **1985**, 415–420.
11. Acker, L. (1968). *Biochemical and Microbiological Aspects of Low Water Activities in Dehydrated Foods*. Institut fur Lebensmittelchemie der Universitat Munster, West Germany.
12. Rossell, J. B. (1983). Measurement of Rancidity, Rancidity in Foods, Applied Science, p. 24.
13. Hadorn, H., Keme, T., Kleinert, J., Messerli, M. and Zurcher, K. (1977). *Chocolate Conf. Bakery Rev.*, **2**(2), 25–38.
14. Worthington and Holley. *J. Am. Oil Chem. Soc.*, **44**, 515–516.
15. Halton, P., Knight, R. A., Martin, H. F. and Ottaway, F. J., (1959). *J. Sci. Food Agric.*, 10 July.
16. Dagnell, Holgate and Reid. Leatherhead Food R.A. Report 642.

Further reading

Hadorn, H., Keme, T., Kleinert, J., Meserli, M. and Zurcher, K. (1972). Behaviour of hazelnuts under different storage conditions. *Chocolate Conf. Bakery Rev.*, **2**(2), 25–39.
Lee, D. M. (1968). *Effect of Light on Foodstuffs*. Leatherhead Food R.A. Literature Survey No. 4, October, 1968.
Leatherhead Food R.A. (1978). *FIRA/Astell Apparatus for Oxygen Absorption by Oils and Fats*, Technical Circular No. 654, May.
Loders & Nucoline Information Bulletins, Unilever.
Kocher, S. P. and Meara, M. L. (1974). *Lipolytic Rancidity in Foodstuffs*, Leatherhead Food R.A. Research Bulletin, 208.

Appendix

Schaal test—a historical note

Extract from the letter written by Schaal, 26.6.1931, which refers to the conditions of the test named after him. (See Rossell, J. B. (1983) *Rancidity in Foods*, Applied Science, p. 35.)

'The best safeguard we have here in the selection of coconut fats or any other shortenings, is to examine each delivery carefully. Among the several tests applied is the one of keeping property. This test has been developed here in the Technical Institute and consists of holding a sample of the fat in an electrically heated cabinet at 145 °F to note the time at which definite rancidity can be detected by odor and taste. This test has been constantly used in this laboratory for three years, not only on coconut fats but on lards, oleo oils, and all other types of shortenings. We have set up specifications in 'days of keeping' on each class. For example, a lard must keep at least 12 days, an all-hydrogenated fat 25 days and a No. 1 oleo oil 12 days.'

10 Rancidity in dairy products
J. C. ALLEN

10.1 Introduction

It is nowadays easier to define 'rancidity' than 'dairy product'. Dairy products have, for some years, included not just the traditional milk, cream, butter and cheese, but yoghurt, ultra high temperature (UHT) pasteurized milk, skimmed and semi-skimmed milk, spray-dried milk, and skimmed milk plus vegetable fats as a whole milk substitute or as specially formulated babymilks. In addition, there is wide usage of dairy ingredients in other foods, such as casein as a meat extender.

Dairy products, like other water-plus-oil emulsions, can suffer from both hydrolytic and oxidative rancidity. Dried products are less susceptible to hydrolysis, but even in these the shelf life is reduced in the presence of atmospheric moisture. The incorporation of only saturated fat into a product removes the problem of oxidative degradation, but this, of course, is not always possible.

10.2 Lipolytic rancidity

Lipolytic rancidity in dairy products would be rare if the milk from which they were made was properly pasteurized or dried. In reality, lipolysis can be a practical problem in liquid and dried milk powder, cream, butter, ice cream and cheese. In milk itself the impairment is detected as a soapy taste and sharp aroma, but in butter and similar products there is a more distinctive off-flavour. Milk triglycerides contain a much higher proportion of shorter chain acyl groups than vegetable or animal depot fats, and the C_4–C_8 fatty acids have their own characteristic tastes. These flavours, due to the action of bacterial enzymes, often develop in dairy products on storage. Lipolysis can also affect the functional properties of dairy products, including reduction of skimming efficiency of raw milk, and a reduction of churning efficiency of cream.

Serious outbreaks of lipolytic rancidity are now rare in the EU. This indicates the considerable progress made since the investigation in 1973–74 of some 3000 butter samples from six European countries. This showed that in two countries some 6% of the samples had free fatty acid

(FFA) levels sufficient to cause flavour impairment, and a further three countries had 20% of samples in this category.

10.2.1 Measurement of lipolysis

Lipolysis is the hydrolysis of triglycerides to produce FFA, which possess undesirable flavour characteristics. Most lipolytic rancidity in dairy products arises from the enzymic action of lipases.

$$\underset{\substack{\text{triglyceride}\\\text{(triacylglycerol)}}}{\begin{array}{l}CH_2OCOR_1\\|\\CHOCOR_2\\|\\CH_2OCOR_3\end{array}} \xrightarrow[\text{lipase}]{H_2O} \underset{\substack{\text{diacylglycerol}\\+\,R_1COOH}}{\begin{array}{l}CH_2OH\\|\\CHOCOR_2\\|\\CH_2OCOR_3\end{array}} \xrightarrow[\text{lipase}]{2H_2O} \underset{\substack{\text{glycerol}\\+\,R_2COOH + R_3COOH}}{\begin{array}{l}CH_2OH\\|\\CHOH\\|\\CH_2OH\end{array}}$$

Lipolysis is usually determined by measurement of the FFA in the fat of a sample. This is achieved by extracting the fat, dissolving it in a neutralised solvent and carrying out a titration. The International Dairy Federation (IDF) Standard Method No. 6A (1969) is commonly followed. The repeatability and reproducibility of the test are good.[1]

There are a confusing number of ways of expressing the results of such titrations, including (a) ml of 1N NaOH or KOH required to neutralise 100 g fat, (b) meq FFA/100 g fat or litre of milk, (c) mg NaOH or KOH (real chaos is likely here!)/g or 100 g fat and (d) mg FFA expressed as oleic acid/100 g fat. The author urges standardisation of units, preferably by the adoption of the chemically rational (b), which is equivalent to mmol titratable H^+ per 100 g fat, or per litre of milk where the exact fat content might not be known. For details of a general method, see Pearson's *The Chemical Analysis of Foods*.

An alternative test for lipolysis is to measure lipase activity itself rather than the FFA products, using indoxyl acetate as a substrate. Atmospheric oxygen oxidises the indoxyl ion which is formed to a blue compound. Although the method is simple, it is not always reliable; in particular, it gives no information on the history of lipolysis in a sample, but only assays its lipolytic potential. Nevertheless, in certain circumstances, this knowledge in itself can be exceedingly useful.

10.2.2 Lipolysis in milk

As little as 1–1.5% lipolysis (i.e. 1–1.5 meq FFA/100 g fat) can render a milk unpalatable: the flavour thresholds of butyric acid and hexanoic acids in milk are only 25 and 14 ppm, respectively. FFA levels in milk are commonly measured by the Bureau of Dairy Industry (BDI) method: fresh milk has an acid degree value (ADV) of 0.5–1 meq/100 g fat and ADVs above 2 indicate unacceptable lipolysed flavour.

The lipases responsible stem either from the milk itself or from bacterial contamination. In milk with total viable counts $< 10^6$ ml^{-1}, intrinsic milk lipoprotein lipase[2] is the prime cause of hydrolysis in cold storage. In fact, this enzyme is present in such quantity in normal milk as to be potentially able to hydrolyse the triglyceride rapidly: calculations indicate its potential activity to be around 2 mol FFA l^{-1} min^{-1}, a rate which would render milk undrinkable within 1–2 min. In reality, the observed rate is only 0.1% of this, being around 0.002 mol FFA l^{-1} min^{-1}.

There are several reasons why the potential activity of milk lipase is not realised in practice. One is that the enzyme occurs in the skim and is occluded by casein micelles, thus being prevented from contact with the fat globules; these themselves are additionally protected by the fat-globule membrane. In addition, there are effective and specific lipoprotein lipase inhibitors in milk, together with a possible lipoprotein activator which resides at the oil–water interface, and which facilitates enzyme–substrate binding.

Since the structure of the milk emulsion physically prevents enzyme–substrate interaction, one might predict that any agitation which would disrupt this structure would result in the sudden onset of lipolysis.[3,4] This is indeed so, and this activity is characterised as induced lipolysis. It can be brought about by any form of agitation, including mixing, stirring, pumping and bubbling. Once induced, the reaction proceeds rapidly, but for a short period only (some 10–15 min) and then, if the milk is quiescent, levels off; further agitations cause successive 'bursts' of lipolysis. The levelling-off occurs because the fat globules, whose membranes have become disrupted, eventually clump together, and perhaps coalesce. Milk lipoprotein lipase requires an oil–water interface for activity, and the clumping markedly reduces the area of interface. Further agitation, and particularly homogenisation, which produces smaller fat globules and hence a larger interfacial area, causes a further period of enzymic activity (Figure 10.1).

Induced lipolysis is thus a potential problem in processing or transportation of milk, whether by pump and pipe or in a tanker. It has been estimated that 50% of the FFA present in pasteurized milk is generated by the time it is pumped into the bulk tank. Keeping pipeline lengths down and ensuring the exclusion of entrapped air assist in minimising induced lipolysis. It is important to consider the design and conditions (air pressure, etc.) of the equipment used,[5] to avoid excessive air intake during milking and to minimise turbulence and agitation.

Spontaneous lipolysis of milk can also occur. This is still due to intrinsic

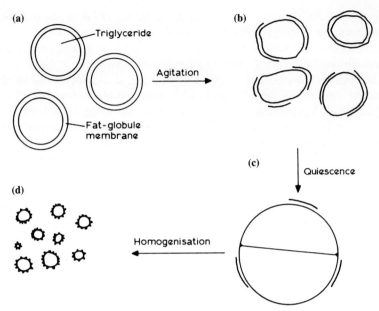

Figure 10.1 Induced lipolysis. (a) Normal milk—little lipolysis. (b) Disruption of globule—lipolysis occurring. (c) Clumping and coagulation—lipolysis ceases. (d) Small globules—rapid lipolysis which soon ceases due to protection by casein and presence of inhibitor.

lipases, but agitation is irrelevant, and the phenomenon, though not fully understood, appears to depend on the individual cow. It is initiated by rapid cooling of fresh milk. Among the factors which have been implicated in its causation are increased levels of blood components in milk, the stage of the lactation cycle of the cow, its nutritional state, a possible previous history of low yield and the intrinsic level of FFA in the milk (these are present in all milks owing to the incomplete synthesis of triglycerides). Chazal and Chillard[6] consider that the seasonal variation in bulk tank FFA is entirely explicable in terms of physiological factors.

There have been a number of investigations on the inhibition of both intrinsic and bacterial lipases in milk by acceptable and normal food components. Some food hydrocolloids such as carageenans have been reported to be effective.[7]

10.2.3 Lipolysis in dairy products

In good quality, microbiologically clean milk, any lipolysis stems from the action of intrinsic milk lipoprotein lipase. The greatest risk of lipolysis occurs prior to heat treatment sufficient to denature milk lipase. In dairy products, high FFA levels can arise either from high levels in the milk used, pre-manufacture lipolysis, or from post-manufacture lipolysis. The latter is caused either by microbial contamination of the product during or after processing, or by contamination with heat-resistant enzymes from psychrotrophic bacteria which are able to grow in milk or cream kept in bulk storage tanks at 2–4 °C. Although the bacteria themselves are inactivated by pasteurization, the proteases and lipases produced by them are well able to survive normal pasteurization temperatures (72 °C for 15–20 s), and Table 10.1 shows that the time required for 90% inactivation of these enzymes is much longer than that normally used in high temperature/short time (HTST) milk production.

It should be clear from the above that considerable care has to be taken to minimise lipolysis in dairy products. When it does occur it is often intermittent and infuriatingly difficult to trace. One example of a problem in cream was due to pin-holes in the pasteurizer, which allowed contaminated water to enter. But the hydrostatic pressure of the water was only more than that of the cream during the brief moments of starting and stopping the machine, so tracking down the fault required a considerable time.

A further illustration concerns a 36% butter fat powder, which was made to a specification of < 1.5 meq FFA/100 g fat. Plant A originally produced it, and random checks on batches showed values of 0.2–0.5 meq FFA/100 g fat, well within specification. Production of this powder was transferred to plant B, and the levels rose to 1.5–2.0 meq FFA/100 g fat. Even if the FFA level was satisfactory immediately after drying, it was found to climb over the limit on storage. Indoxyl acetate tests for lipase activity proved negative, but a sensitive titrimetric method using a pH-stat showed some activity to be present. Such post-manufacture lipolysis was due to heat-resistant microbial lipases, and was eliminated by improving the quality of raw milk used by plant B.

Sometimes flavour faults can be wrongly ascribed. A dried milk preparation from blended skimmed milk, vegetable oil including a small proportion of sunflower oil, and a small amount of buttermilk, was

Table 10.1 Times required for 90% inactivation of some bacterial enzymes

Organism	Enzyme	Time (min)
Ps. fluorescens 22F	Lipase	4.8
Ps. fragi 14-2	Lipase	0.27
S. marcesens D2	Lipase	0.38
Achromobacter spp 230	Lipase	33
Ps. putrefaciens R48	Lipase	0.74
Pseudomonas 21B	Lipase	170[a]
Pseudomonas 21B	Proteinase	250[a]

[a]Calculated and/or extrapolated

prepared by a company using its own dairy components but with bought-in vegetable oils. A batch of the dried product developed a taint after six weeks' storage, and the defect was considered to be oxidative, probably from the sunflower oil. Titration of FFA had shown that 'lipolysis could not be the cause; anyway, all our skim is our own and of very high quality.'

The author's view on being asked to investigate two weeks later was that the flavour had a soapy note and the problem was possibly lipolytic. Peroxide and anisidine values were low, and a requested indoxyl acetate test showed lipases to be present. On further investigation the culprit was shown to have been probably the small proportion of buttermilk used in the formulation.

In butter, the type of cream employed, the salt concentration and the pH can affect both the actual and perceived lipolytic rancidity. For instance, ripened cream butter has a 'lactic' flavour anyway, so higher levels of FFA can be tolerated than in a sweet cream butter. In general, a low pH (4.8) and high salt concentration (2%) is inhibitory to microbial growth, but at higher pH (6.5) with low added salt, the primary cause of flavour impairment is lipolysis. In some cheeses a lipolytic flavour is not only acceptable but desirable.

The International Dairy Federation has published typical ranges of normal and unacceptably high FFA-values in dairy products.[8] These guide values are given in Table 10.2.

10.3 Oxidative rancidity

Unacceptable oxidative rancidity is more common in dairy products than lipolytic rancidity, but is less common in milk itself, partly since milk is not usually kept for so long before use as other dairy products. Autoxidation is less predictable than lipolysis: although many of the factors

Table 10.2 FFA concentrations in dairy products; rancid flavour threshold values

Product	FFA-values (meq/100 g fat)	
	Normal values	Likely to cause problems
Milk: milk powder	0.3–1.0	1.5–2.0
Cream: ice cream	0.5–1.2	1.7–2.1
Butter	0.5–1.0	1.0–2.1
Cheese: Cheddar	1.2	2.9
Brie	1.2	–
Blue	40	–

involved are known, they affect the reaction when present in only trace amounts (see chapter 1 for a description of the chemistry of oxidative rancidity).

10.3.1 Measurement of oxidative rancidity

Chapter 2 deals with measurement of lipid oxidation in detail and in a general context. In milk and dairy products, assessment of lipid oxidation usually implies determination of the peroxide value (PV), normally expressed in meq/100 g or kg fat, the fat having been de-emulsified previously. A method of de-emulsification of milk powders (which can be adapted for milk) found satisfactory by our group is based on that of Pont.[9] Sodium citrate (10 g), sodium salicylate (10 g) and butan-1-ol (18 ml; analytical grade) are added to a conical flask fitted with a glass stopper. Water (72 ml) is added, and the reagents mixed by vigorous shaking. The reagent should be used within 45 min of preparation. Water (30 ml) and milk powder (10 g) are added to another similar conical flask and the contents shaken. The de-emulsification reagent (15 ml) is added and the flask again shaken. It is placed in a water bath at 65–70 °C for 10 min, centrifuged for 5 min and replaced in the water bath for 5 min. The fat is removed with a Pasteur pipette and directly weighed into the peroxide value determination flask.

The peroxide value is determined in a flask protected as far as possible from light by aluminium foil. The sample, chloroform (10 ml) and glacial acetic acid (15 ml) are added, and freshly prepared saturated aqueous KI (1 ml). The contents are gently swirled for 1 min and water (75 ml) is added. The solution is titrated against freshly prepared 0.002 M $Na_2S_2O_3$ using starch as an indicator.

A 'blank' titration is always performed and the results deducted from the sample titration. If the 'blank' value exceeds 0.5 ml of 0.002 M $Na_2S_2O_3$, the reagents are discarded.

Other more sensitive fluorescent[10] and enzyme-linked[11] assays have been described, which are said to correlate well with the results from titration.

10.3.2 Oxidative rancidity in milk

Provided contact with copper is rigorously avoided, and use of modern dairy equipment should ensure this, there is no general problem of oxidative rancidity in pasteurized milk itself, at least over 14 days from pasteurization. Naturally, exceptions occur when the milk is mistreated: light-induced oxidation can easily be induced in milk left on sunny doorsteps, giving rise to an off-flavour described as 'cardboard-like'.

Although unacceptable oxidative rancidity in pasteurized milk is rare, work from our own laboratory has indicated that a characteristic of pasteurized milk described as 'stale' may be oxidised flavour. Freshly pasteurized milk was stored at 7 °C in the dark, in sterile glass bottles with no headspace. Bottles were opened at intervals, and their contents chemically examined and organoleptically assessed. The taste panel was asked to determine whether the milk was fresh or 'stale' (which was operationally defined as a flavour different from that of a fresh control sample) and if 'stale' to report whether the change was most similar to oxidised, lipolysed, light-induced, microbial or heated flavours. By far the most frequent term used was 'oxidised'.[12]

By contrast, the current widespread use of UHT milk, with its much longer storage capability, can give rise to a staleness which might well be due to hydrolytic or oxidative rancidity. An attempt has been made to establish criteria to provide a basis for uniform legislation on the 'best before' date for UHT milk in the EC.[13]

Bassette's group, working with UHT milk, have found that, although the content of dissolved O_2 affected the rate of stale flavour development, saturated aldehydes were not principal contributors to the flavour.[14] Changes in ADV correlated best with stale flavour, and levels of short-chain aldehydes and ketones were below their flavour thresholds.[15] It is, however, difficult to correlate the results from different studies in this area. Different UHT treatments, storage conditions and packaging can give different and often apparently irreconcilable findings. For instance, quite high levels of carbonyl compounds can be absorbed into packaging from UHT milk or cream.[16]

Milk from some herds is remarkably resistant to oxidation, whereas milk from other herds can develop oxidised off-flavours for no apparent reason: such spontaneous oxidation needs further research to clarify the issues involved. Since it is more prevalent in winter months, it has often been associated with dry feed. However, other factors, such as feedstuffs containing oxidised fat, copper-contamination of the water supply and excessive agitation have also been implicated. Further research on spontaneous oxidation of milk is necessary to elucidate causes and suggest means of minimisation.

The triglycerides of milk fat contain levels of polyunsaturated acyl groups which, if they were substantially oxidised, would be more than sufficient to give rise to totally unacceptable flavour impairment. Milk also contains potential catalysts of lipid oxidation, in the form of native Cu^{2+} and metalloenzymes like xanthine oxidase and lactoperoxidase. Since milk is resistant to oxidative deterioration, it seems reasonable to expect the existence of antioxidants in milk. Note, incidentally, that there is an interesting parallel here, in that milk has far more potential for both lipolytic and oxidative rancidity than is actually observed.

Ascorbic acid, whose concentration in freshly pasteurized milk is 15.8 ± 2.8 (standard deviation, $n = 28$) mg l^{-1},[17] appears to act as a lipid antioxidant. Its concentration decreases steadily during storage, and reaches zero by a week from pasteurization, having consumed about 33% of the dissolved O_2 in the process. Tocopherols are also present in milk, and possibly also assist in this antioxidant role. The peroxide value of good pasteurized milk stored at 7 °C is well below 5 meq/kg fat for the first week, but the problem is greater in UHT milk because of its longer shelf life and storage temperature. The ascorbic acid disappears in UHT milk just as quickly as in pasteurized milk, and it is then more susceptible to copper- or light-induced off-flavour.[18]

However, milk is protected from lipid autoxidation by means other than chemically conventional antioxidants. The major proteins, casein especially, are quite effective at binding metals such as Cu^{2+} and Fe^{3+}, and this can physically prevent contact between the metal ion and its triglyceride substrate. In addition, raw (and, to a lesser extent, pasteurized) milk has its triglyceride protected by the fat-globule membrane. Although this contains metalloenzymes such as xanthine oxidase, they also appear to be held away from the triglyceride core of the fat globule, and do not induce significant lipid oxidation provided the membrane is not disrupted. Homogenization of milk disrupts the membrane but leaves the fat droplets coated with casein micelles, thus maintaining the emulsion. This is often regarded as conferring the bonus of protecting the fat from oxidation, but there appears to be little evidence for this. In our experience, the fat in homogenized milk is somewhat more susceptible to oxidation, and more recent work[19] gives evidence for the enhancement of Cu^{2+}-catalysed oxidation in milk fat in the presence of milk fat-globule membrane material.

10.3.3 Oxidative rancidity in dairy products

In milk and dairy products, Hill's group and our own group have shown that pasteurization at 80 °C can be an effective means of reducing the pro-oxidative effects of lactoperoxidase and xanthine oxidase.[20,21] The latter can act as a pro-oxidant in the absence of added Cu^{2+}, but even small amounts of Cu^{2+} can cause a striking increase in its effectiveness, at the same time as abolishing its normal enzymic activity. Hill *et al.* have shown that superoxide dismutase (plus catalase, to remove the H_2O_2 formed) reduces development of oxidised flavour in heat-treated 'high linoleic acid' milk in the absence of added Cu^{2+}, but is ineffective when 0.1 ppm Cu^{2+} is added after heat treatment. The superoxide dismutase was thus effective at removing any superoxide anion ($O_2^- \cdot$) generated by the native milk xanthine oxidase, but when Cu^{2+} was added the $O_2^- \cdot$ was presumably converted to the very reactive hydroxyl radical ($\cdot OH$).

Processing milk does not remove the factors which render the fat liable to oxidation. On the contrary, separating, churning, drying and additives can often increase lipid oxidation, by the intermixing of catalysts with the lipid and by contamination from the plant. Migration of Cu^{2+} into the cream on churning can cause rapid flavour impairment. Buttermilk is often highly susceptible to lipid oxidation: its high proportion of unsaturated phospholipids (in particular, phosphatidylethanolamine) can bind metal ions in a pro-oxidative fashion, and the presence of a metal–phospholipid complex at an oil–water interface can facilitate lipid hydroperoxide formation.

Butter itself can oxidise, both by metal ion catalysis, which gives rise to 'fishy' taints, and by photo-oxidation, which is often due to non-foil-wrapped butter being stored near to the fluorescent light of a display chill cabinet.[22] Dried whole milk is usually fairly resistant to oxidative rancidity provided it is kept dry. However, 'humanised' babymilks can give serious problems: they contain relatively high proportions of polyunsaturated fats, and normally have extra Cu^{2+} added for (dubious) nutritional reasons. Careful control has to be exercised over the order of addition, the temperature and duration of blending and drying, and the type of packing. It is both illegal and imprudent to add antioxidants to such babymilks.

There are a number of natural antioxidants which occur in various foodstuffs which can protect butter-based products from oxidation. One finding with possible application is the effect of Maillard reaction products, which were more effective at 500 ppm than butylated hydroxytoluene (BHT) in protecting butter oil.[23]

A number of groups have endeavoured to find ways of adding nutritionally desirable trace metals to milk and milk products in a form which would not be pro-oxidative. Usually, the method is to add the metal as a complex, and the metabolically acceptable citrate or lactate has been used. However, no really satisfactory protection has normally been achieved, and the use of stronger complexing agents such as EDTA or nitrilodiacetate is questionable in, for instance, infant feeds. Certain gums, such as the carageenans, have protective properties against lipid oxidation in foods, and their effects need to be explored further.

An alternative approach is to add the Cu^{2+} in protective microcapsules or entrapped in denatured casein. However, the only sure way to protect a very susceptible product is to gas-pack under nitrogen.

Acknowledgements

My thanks are due to several colleagues in the dairy industry for (unattributable) examples of problems of rancidity.

References

1. Jellema, A., Oger, R. and Van Reusel, A. (1988). *Bull. Int. Dairy Fed.*, **235**, 81–91.
2. Yang, C. Y., Gu, Z. W., Yang, H. X., Rohde, M. F., Gotto, A. M. and Pownall, H. J. (1989). *J. Biol. Chem.*, **264**, 16822–7.
3. Cartier, P. and Chillard, Y. (1989). *J. Dairy Res.*, **56**, 699–709.
4. Lehmann, H. R. (1988). *Dtsch. Molk Ztg.*, **109**, 634–636.
5. Needs, E. C., Anderson, M. and Morant, S. V. (1986). *J. Dairy Res.*, **53**, 203–10.
6. Chazal, M. P. and Chillard, Y. (1986), *J. Dairy Res.*, **53**, 529–38.
7. Stern, K. K., Foegeding, E. A. and Hansen, A. P. (1988) *J. Dairy Sci.*, **71**, 41–45.
8. International Dairy Federation (1987).*Significance of Lipolysis in the Manufacture and Storage of Dairy Products*, Annual Sessions in Helsinki, Report of Group B33, Document 144.
9. Pont, E. G. (1955). *Austral. J. Dairy Technol.*, **10**, 72.
10. Akasaka, K., Sasaki, I., Ohrui, H. and Meguro, H. (1992). *Biosci. Biotechnol. Biochem.*, **56**, 605–607.
11. Akasa, I. and Aota, N. (1990). *Talanta*, **37**, 925–929.
12. Allen, J. C. and Joseph, G. (1983). *J. Dairy Technol.*, **36**, 21.
13. Glaeser, H. (1989). *Dtsch. Molk Ztg.*, **110**, 580–585.
14. Wadsworth, K. and Bassette, R. (1985). *J. Food Protect.*, **48**, 487–93.
15. Jeon, L. J. and Bassette, R. (1987). *J. Dairy Sci.*, **70**, 2046–2054.
16. Hansen, A. P. and Arora, D. K. (1990). *Am. Chem. Soc. Symp. Ser.*, **423**, 318–332.
17. Allen, J. C. and Joseph, G. (1983). *J. Dairy Res.*, **52**, 469–87.
18. Schroder, M. J. A. (1982). *J. Dairy Res.*, **49**, 407.
19. Nawar, W. W. (1991). *J. Food Sci.*, **56**, 398–401, 446.
20. Hill, R. D., van Leeuwen, V. and Wilkinson, R. A. (1977). *New Zealand J. Dairy Sci. Technol.*, **12**, 69.
21. Allen, J. C. and Wrieden, W. L. (1982). *J. Dairy Res.*, **49**, 249.
22. Emmons, D. B. *et al.* (1986). *J. Dairy Sci.*, **69**, 2437–50.
23. Farag, R. S., El-Baroty, G. S. and Hassan, M. N. A. (1989). *Chem. Mikrobiol. Technol. Lebensm.*, **12**, 111–118.

Further reading

Lipolytic rancidity

International Dairy Federation (1975). *Proceedings of the Lipolysis Symposium*, Cork, Ireland, Document 86.
International Dairy Federation (1979). *Flavour Impairment of Milk and Milk Products due to Lipolysis*. Report of Commission A, Document 43.
Pearson, D. (1976). *The Chemical Analysis of Foods*, 7th edn., Churchill-Livingstone, Edinburgh, p. 493.
Driessen, F. M. (1989). *Bull. Int. Dairy Fed.*, **238**, 71–93. A review of milk lipases and proteases.
Olivecrona, T. and Bengtsson-Olivecrona, G. (1991). In *Food Enzymology* (ed. P. F. Fox) Vol. 1, Elsevier, London, pp. 62–78. A review of milk lipoprotein lipase.
Soerhang, T. and Stepaniak, K. (1991). In *Food Enzymology* (ed. P. F. Fox) Vol. 1, Elsevier, London, pp. 169–218. A review of microbial enzymes in milk and dairy products.
Seitz, E. W. (1990). *J. Dairy Sci.*, **73**, 3664–91. Review of microbial and enzyme-induced flavours.

Oxidative rancidity

Lundberg, W. O. (ed.) (1961). *Autoxidation and Antioxidants*, Interscience, New York.
Schultz, H. W., Day, E. A. and Sinnhuber, R. O. (eds.) (1962). *Symposium on Foods: Lipids and their Oxidation*, Avi Publishing, Westport, CT., USA.
Simic, M. G. and Karel, M. (1980). *Autoxidation in Food and Biological Systems*, Plenum Press, New York.

11 Rancidity in meats
M. D. RANKEN

11.1 Introduction

The topic of rancidity in meat and meat products has been well reviewed by a number of authors.[1-4] The chemical principles at work are the same as in other foods, but their effects are modified in a number of ways, sometimes quite drastically, by factors peculiar to meat.

11.2 Features special to meat

The most important of these factors is the structure of the meat fatty tissues. What the butcher or meat manufacturer calls 'fat' is not the refined or rendered fat familiar to a margarine manufacturer or many research chemists, but is unrendered adipose tissue. In this tissue the lipids are enclosed in small cells (diameter ca. 1 μm) made of connective tissue, massed together in the familiar honeycomb formation (Figure 11.1).

The cell walls consist mainly of collagen, with other proteins, hydrated with ca. 75% water. Access of any added substance to the lipid inside the cells is hindered mechanically by this cellular structure. For any chemical

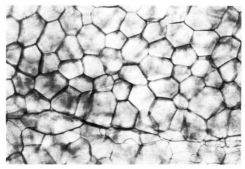

Figure 11.1 Lipids enclosed in small cells in meat tissue. Courtesy of the University of Bristol, Division of Food Animal Science.

reaction to take place between the lipids and other substances not normally present in the fat cells, the other substances must first be transported from outside the meat, in the aqueous phase, along and through the fatty tissue cell walls. The reactions with lipid must then take place not in the aqueous phase but in the lipid phase or at the interface. Not many of the substances with the potential to be involved in lipid reactions can easily satisfy these solubility requirements.

When meat, including its fat, is minced or chopped, as happens in making many meat products, the cellular tissue is more or less broken down so that the lipids are exposed to chemical attack from which they were previously protected. At the same time the muscle tissue is also broken down and its constituents made more readily available for chemical or microbiological interactions. Chemical changes including rancidity are therefore more likely to occur, or to proceed more rapidly, in comminuted meat and meat products than in the same meat without comminution.

In the intact fat cell there may be natural antioxidant systems present. Disruption of the cells not only exposes the internal lipids to chemical attack but may also inactivate any protective antioxidants which may have been present.

11.3 Types of rancidity

11.3.1 Hydrolytic rancidity

Hydrolytic rancidity is only significant in meat and meat products in certain circumstances. The hydrolysis may be due either to direct chemical causes or to enzyme activity. The commonest chemical cause in other fatty foods is the acidity of the food, either in its natural state or as developed in the course of other deteriorative changes. However, in meat the natural pH value is close to neutrality, 7.0 to 7.2 in the live animal, falling to 5.5 to 6.5 after death and only very exceptionally as low as 5.0. Muscular tissue has a high buffering capacity and if acids are added to it, quite large quantities of strong acid are necessary to alter the pH value significantly. Furthermore, because of the structural factors outlined above, the possibility of chemical reaction between the lipids in the fatty tissue and acids in the muscular tissue is small, so hydrolytic rancidity in ordinary meat is not very common.

Hydrolytic changes initiated by enzymes may occur in meat or meat fats where there is microbiological growth. Lipolytic enzymes produced by micro-organisms on the meat surface, especially by mould but sometimes by yeasts or bacteria, may well produce rancidity in the surface fat.[5] In such circumstances microbially-induced proteolytic changes usually also occur. Proteolysis damages the fatty tissue cell walls and makes the fat more accessible to the lipolytic enzymes. However, the general level of microbiological activity usually leads to such obvious mouldiness, slime or smell that the meat is considered spoiled quite independently of any increase in rancidity. The problems can normally be avoided by minimising microbial contamination, by good general hygiene and by storage at low temperatures.

In certain dried fermented sausages the chemical and enzymic conditions are controlled so as to encourage some development of hydrolytic rancidity. In the process of manufacture lactic acid is produced, bringing the pH value of the sausages down to 5.0 to 5.5, and in the later stages of the process, when conditions are favourable, mould is introduced and allowed to grow on the surface. The rancidity thus developed is considered to be an essential component of the characteristic flavour of the products.[6]

11.3.2 Oxidative rancidity

Oxidative rancidity is much more common in meats than hydrolytic rancidity. Note again that the general chemical principles involved apply equally in meat systems as in other foods, and although there may be modifications in their effects due to the nature of meat structure and other causes, some important simple predictions can be made from basic principles.

First, it is fundamental that the more unsaturated the fat, the more it is prone to oxidation: rancidity therefore develops faster and further in the relatively unsaturated pork fats than in the harder beef or mutton fats, faster again (other factors being equal) in the very soft chicken fat and fastest of all in fish oils.

As well as these well known species differences in fat hardness, in the animal body the lipids directly around the internal organs are harder (with higher melting points and lower degrees of unsaturation) than those on the outside, and, likewise, the inner layers of body fat are harder than the outer layers.[7] These differences may be related to a need in the live animal for lower melting points in the fats further away from the body centre, where the temperatures are slightly lower, so that the lipids in the tissue cells may always be in the liquid state.

(In pork, the cell walls of the fatty tissue are thinner in the tissues which contain the more saturated and therefore harder, lipids. This has the curious consequence that if hardness is assessed manually when the animal or the fatty tissue removed from it is still warm, tissues with relatively saturated lipid but thin cell walls will feel softer than those containing more unsaturated lipids. At lower temperatures when the hardness of the tissue is governed by the proportion of crystallised lipid, comparison by manual assessment gives the opposite result.)

There are differences in susceptibility to oxidation among different cuts of meat or individual muscles.[8] These are probably related to differences in degree of unsaturation of the fats present but other factors may also be involved.

The second principle is that for oxidation to occur an oxidising agent is required and this must be able to gain access to the fat. The commonest oxidising agent, of course, is oxygen from the air. Measures which exclude oxygen, such as vacuum packaging, or which reduce its concentration, as in modified atmosphere packaging, may be expected therefore to prevent or reduce oxidative rancidity. On the other hand, comminution of the meat and the resultant increased exposure of the lipids will increase it.

In the case of the phospholipids in meat fat both these factors are at work. The phospholipids contain highly unsaturated fatty acids, very prone to oxidation, and they are located predominantly in or at the cell walls of the fatty tissue. They are therefore the first to be exposed to oxidation when the cells are damaged. Other phospholipids are more or less finely dispersed in the lean meat or musculature, where similar considerations apply.

11.4 Special factors in meat which affect oxidative rancidity

11.4.1 Meat pigments

There are important connections between lipid oxidation in meat fats and oxidation of the meat pigment myoglobin, present in the aqueous phase of the muscle tissues.

11.4.1.1 Chemistry of myoglobin. Myoglobin is a water-soluble compound of a protein (globin) with a pyrrole complex called haem. Haem contains an iron atom which is readily subject to changes in its oxidation state between Fe^{II} and Fe^{III}. The physiological activity of myoglobin in live muscle consists principally in the transmission of oxygen from the bloodstream (where oxygen is carried by haemoglobin) to the sites of muscle activity within the muscles themselves. In meat, after the death of the muscle, this oxygen-carrying property of myoglobin remains important for the maintenance of a 'fresh' red colour. (The chemistry of the blood pigment haemoglobin is similar to that of myoglobin: however, since little free blood remains in butcher's meat, and what there is tends to be retained in the blood vessels, the concern here is mainly with myoglobin.)

Myoglobin can carry or release oxygen only when the iron of the haem is in the Fe^{II} state. Such a myoglobin molecule may exist either in an oxygenated form, loosely bound to one molecule of oxygen, or in an unoxygenated form without any oxygen molecule. The colour in the former state is the familiar bright cherry-red of fresh meat; in the latter, it is a deep purple-red. (Similar colour differences are well known between arterial blood, carrying oxygen, and venous blood with none.) If the iron becomes oxidised to Fe^{III}, the haem complex becomes insoluble, the colour changes to brown and no oxygen can be bound. This oxidised form is known as met-myoglobin.

Oxidation of myoglobin to met-myoglobin occurs at low oxygen concentration, corresponding to about 3 mm partial pressure of O_2.[9] It is reversible on change of oxygen concentration, through the action of enzymes normally present in meat, but only very slowly: for practical purposes therefore met-myoglobin, once formed in the meat, must be regarded as fairly stable.

The oxygenated form is easily produced in the presence of much oxygen and converts to the unoxygenated form ('myoglobin' or 'reduced myoglobin') when oxygen is absent: these reactions are readily reversible on the appropriate changes in oxygen concentration.

These relationships are described in Figure 11.2. For more detail the reviews by Ledward[10] and Renerre[11] may be consulted.

The practical consequence of these chemical changes is that so long as abundant oxygen is available at the meat surface, the colour of that surface remains the desirable bright red of oxy-myoglobin. Complete removal of oxygen, as in vacuum packing of the meat, converts the myoglobin to its reduced, unoxygenated form, which can be converted back to bright red oxy-myoglobin as soon as the package is opened and the meat exposed to oxygen again.

In meat or meat products which are not vacuum packed, other factors are at work which consume oxygen and remove it relatively slowly from the system. The most important are the residual metabolic activity of the muscles themselves and the metabolism of any micro-organisms which may be present. The former operates within pieces of meat, but the latter tends to operate only at the surface where micro-organisms are located, except when the meat is comminuted and the organisms become redis-

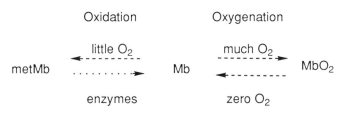

Figure 11.2 Oxidation and oxygenation of myoglobin.

tributed throughout the mass; they can then grow and consume oxygen within the mass of the meat as well as at the surface. As this happens, oxygen becomes depleted at the centre while remaining available in the air at the surface, so a concentration gradient is set up between the centre and the outside. Along that gradient, at a certain distance below the surface, there will be a zone of low oxygen concentration where myoglobin becomes oxidised to met-myoglobin, forming a dull brown insoluble layer beneath the transparent red surface. The meat is said to have lost its 'bloom'. Eventually the met-myoglobin layer becomes so thick that the meat appears brown and stale.

The colour of cured meat products such as bacon, ham, frankfurters or salami is due to nitroso myoglobin. This occurs in a number of forms depending on the oxidation state of the haem iron and whether or not the globin is denatured. These pigments do not bind oxygen and do not exist in oxygenated forms. The protein-denatured versions, present in cooked cured meats, are readily destroyed by oxidation, especially on exposure to light, which favours free radical formation.

In modified atmosphere packing, fresh meat or meat products are packed under high concentrations of carbon dioxide which suppresses both oxidative rancidity and microbiological growth. However, with fresh meat it is necessary to ensure that the concentration of oxygen in the gas mixture in contact with the meat does not fall below 10% to avoid browning due to met-myoglobin formation.[12]

11.4.1.2 Interaction with fat oxidation. In practice, oxidative changes almost always lead both to rancidity in the fat and to brown discoloration of the lean meat. Brownness in a product such as an uncooked hamburger may very often be a prior indication of an incipient rancid or stale flavour in the cooked product. It is also commonly observed that the use of oxidised fat in a manufactured product promotes accelerated browning, and similarly that fats smeared with blood or meat juices are more prone to rancidity.

Early investigators[13] considered the two oxidation reactions to be synergistic or autocatalytic. More recently, Ledward and his co-workers suggested that the reactions are in fact independent even when they occur simultaneously, but that the fat peroxidation reactions may be catalysed by free iron derived from haem or other sources.[14] The influence of fat oxidation on the colour fading of cured meats is of less importance than with fresh or uncured myoglobin colours.

11.4.2 Effect of freezing

Fat oxidation in meat is one of the few chemical reactions to be accelerated at temperatures below the freezing point of water,[15] with a maximum rate in the region of −10 °C. The freezing points of animal lipids are normally just below the body temperature of the live animal, say around 35 to 40 °C, and much higher than the temperature of commercially frozen food products. However, the acceleration of rancidity with reduction of temperature is related to the freezing not of the lipid but of the water in the fatty tissue. As the temperature is progressively reduced below 0 °C, an increasing proportion of pure ice is formed, whilst the remaining liquid water forms a solution of increasing concentration and lowered freezing point; 98.2% of the water is frozen at −20 °C[16] and not until the temperature reaches ca. −30 °C is all of it solidified. Although the rate of a chemical reaction diminishes with falling temperature, in the case of peroxidation of fats in the range 0 to −10 °C this deceleration is more than offset by the accelerative effect of the increased concentration of the reactants.

The storage life of properly wrapped frozen meat is usually terminated by the onset of rancid flavours. Some recommendations concerning shelf life are given in Table 11.1.

Note the shorter expected shelf life of the meats with more unsaturated fats, also the shorter shelf life of cut or ground products, including sausages. Hygiene in the preparation of products for storage, integrity of the packaging and the reliability of temperature control, all affect the storage life of frozen meats. It is probable that all these, as practised in the meat trade, have been improved since the recommendations in Table 11.1 were made, for satisfactory storage life under commercial conditions may now often be longer than the table shows.

Frozen pork fat, with its relatively high proportion of unsaturated fatty acids, is notoriously prone to rancidity after prolonged storage. Use of such material in manufactured products, even when the rancidity is incipient and not yet overt, is a frequent cause of poor colour, for the reasons previously discussed.

Table 11.1 Storage life of some meats and meat products at −18 °C (International Institute of Refrigeration, 1964)

Product	Expected storage life (months)
Beef	8–12
Roasts, steaks; packaged	12
Ground meat; packaged, unsalted	4–8
Veal, roasts, chops	8–10
cutlets, cubes	6–8
Lamb	6–10
Pork	4–6
ground, sausages	3–4
Poultry (moisture proof packs)	6–8
fried chicken	3–4

11.4.3 Effect of salt

Peroxidation is accelerated in the presence of salt[17] and is greatly accelerated when salted meat is frozen.[18] Frozen storage of cured meats such as bacon must therefore be regarded as hazardous unless the meat is very reliably vacuum packed and the packages carefully handled to avoid damage and exclude all possibilities of access of oxygen.

11.4.4 Rancidity and nitrosamine formation

Great concern was expressed in the 1970s over the discovery that nitrosamines could be formed by the nitrites present in cured meats, with the possibility of a link between such nitrosamines and the development of cancers in humans.

On cooking cured meats at temperatures of 80 °C (frankfurters) to 175 °C (bacon) minute quantities of nitrosamines are formed. The chemistry of their formation is complicated, but the components of the reaction are now considered to include peroxidised lipid, protein from the adipose connective tissue and nitrite present in the aqueous phase.[19,20]

Since oxidised fat takes part in the reaction, antioxidants may be used to prevent or restrain it. α-Tocopherol added to the product during manufacture diminishes (but does not eliminate) nitrosamine formation. Moderate concentrations of ascorbic or erythorbic acid (470 mg kg^{-1}, or 550 mg kg^{-1} as the sodium salts) are similarly effective and are now legally required to be added to bacon in the USA.

The influence of ascorbates on nitrosamine formation is also very complex. As well as its antioxidant effect, ascorbate reduces the concentration of nitrite in the cured meat and as the residual nitrite content is reduced the microbiological risks, most notably that from *Clostridium botulinum*, are increased. The use of ascorbates for control of nitrosamines therefore requires very careful control and excessive concentration must be avoided.

11.5 Rancidity and meat flavour

The relationship of rancidity to the flavour of meat has been reviewed by Mottram.[21] The flavours of fatty acids resulting from hydrolytic rancidity are not necessarily offensive. Lea[22] considered the flavours of the meat fatty acids to be less objectionable than those from milk fats and the contribution of fatty acids derived from hydrolysed fat to the flavour of dried fermented sausages has already been mentioned.

When meat is cooked the lipids are oxidised. The range of breakdown products from the unsaturated fats, mainly aldehydes, alcohols, furans and hydrocarbons, is broadly similar to that produced by oxidation at lower temperatures. But the relative concentrations are not the same and the profile of the volatiles produced is further modified by the oxidation also of the saturated fats when heated. Any hydroperoxides formed on cooking are readily decomposed at cooking temperatures. The end-result is a desirable flavour in cooked meat compared with the undesirable one when fat goes rancid at lower temperatures. However, if off-flavours were already present in the meat before cooking they tend to persist afterwards.

The Maillard reaction between proteins and reducing sugars is well known as a contributor to the flavour of roasted meat. Recent work has demonstrated an essential involvement of lipid in the formation of these flavours, with the suggestion that the highly unsaturated phospholipids may be particularly important.[23]

'Warmed over flavour' may develop in cooked meats stored at refrigerator temperatures for 1 to 2 days and then reheated.[24] Recent work has now shown that this is due to autoxidation of the highly unsaturated phospholipids in the meat, in particular those with four or more double bonds. The changes are catalysed by iron released from haem and inhibited by natural antioxidants (α-tocopherols) in the meat,[10] but the reactions are complex and not yet well understood.[25] Formation of 'warmed over flavour' is also encouraged by the disruption of the cellular structure which occurs on heating, or, of course, on comminution, as happens in the manufacture of many meat products.

It has long been known that 'fishy taints' may be produced in the meat of pigs or poultry fed on diets which contain high proportions of fish oils or fish meals, and the extent of rancidity in the diets has been of some concern in that connection. With beef or sheep meat the problem does not arise because, in them, the unsaturated fats are hydrogenated in the rumen, destroying any previously-formed peroxides as well as producing saturated fat which is less susceptible to oxidation. However, the 'fishy' taints do not always originate from rancid fats. Recent work on fishy taint in eggs has shown this to be due to trimethylamine, produced by microbial spoilage of the protein component of fish meal in the diet of the hen, either before ingestion by the hen or by the flora of the hen's gut after the meal is consumed.[26]

11.6 Control of rancidity

11.6.1 *Use of antioxidants*

Whereas fat-soluble antioxidants are widely used in refined fats and oils and products made from them, they are of little value in fresh or frozen meat cuts or in those meat products which consist of large pieces of meat (e.g. hams), even where the proportion of unsaturated fats is high and rancidity may be troublesome. The reason for this is the prevention of access of any added antioxidant to the lipid inside the fat cells by the cellular structure of the fatty tissue, as previously discussed.

With products made from or containing comminuted meat and fatty tissue antioxidants may be more useful. The addition of ascorbic acid or α-tocopherol to meat patties and sausages for the control of 'warmed over flavour' has been tried, with some success.[27] The proprietary substance Ronoxan D20, a mixture of ascorbyl palmitate with α-tocopherol, was introduced originally to control a condition in British fresh sausages known as 'White Spot'. The cause of this is uncertain but it is apparently associated with a bleaching of the meat pigments at the surface of the sausage, in localised spots where the SO_2 concentration is high. The amphiphilic nature of ascorbyl palmitate permits it to be transported in the aqueous phase through the meat and the fatty tissue cell walls, then into the fat phase to act as an antioxidant.

As well as controlling White Spot, the Ronoxan mixture improves the myoglobin colour stability of the sausages. This is for the reasons outlined above: the formation of met-myoglobin is prevented or reversed by the strong reducing agent, so myoglobin remains available to form bright red oxy-myoglobin instead. Other reducing substances such as ascorbic acid, erythorbic acid or cysteine are also effective: around $200\,\text{mg}\,\text{kg}^{-1}$ of ascorbic acid or an ascorbate will usually extend the shelf life of sausages or beefburgers, before colour fading, by 1–2 days. However, it is not permitted to add them to unprocessed butcher's meat for this purpose. This is because in practice the commonest cause of the loss of the bright colour of fresh meat or an uncured meat product is not rancidity but microbial growth, which consumes oxygen and reduces the concentration of free oxygen below the surface to the partial pressure at which met-myoglobin can be formed. Strong reducing substances prevent this by scavenging the last traces of oxygen, thus preventing met-myoglobin formation. This in effect suppresses the visual evidence of bacterial growth and incipient spoilage, and for this reason the addition of reducing agents to otherwise unprocessed meat is usually prohibited. In the UK and some other countries antioxidants are permitted in meat products, where, although the microbial risk may in theory be higher, it is considered to be under closer control.

Sulphur dioxide and the sulphites are legally permitted preservatives, suppressing microbial activity, and also have the beneficial effects of reducing agents on colour and on fat rancidity.

11.6.2 *Vacuum or modified atmosphere packaging*

The most obvious precaution to take against oxidative deterioration is to remove the source of oxygen. In most cases vacuum packing of meat or meat products, or modified atmosphere packaging in mixtures of carbon dioxide with nitrogen or oxygen, afford very satisfactory protection against colour and rancidity problems. Of course, there are usually also microbiological advantages.

With uncooked cured meats such as unpasteurized bacon the main advantage of vacuum packing is microbiological and the advantage to rancidity or colour is relatively small. With the cooked cured meats, microbiological spoilage is slower and vacuum packing gives significant protection against colour loss and rancid off-flavours.

11.6.3 Avoidance of pro-oxidants

Similar to the removal of oxygen, it is an obvious yet sometimes overlooked precaution, to eliminate or minimise pro-oxidant conditions as far as possible. Substances and conditions which are well known to cause difficulty under practical conditions of use include:

- chlorine and chlorine-based cleaning agents, sanitizers etc.;
- ozone (e.g. from arc welding equipment);
- metal ions, especially iron and copper;
- free radicals—exposure to light, especially ultraviolet light, is highly destructive of the colour of cooked cured meats and in extreme cases may also be the cause of rancid flavours.

References

1. Allen, J. C. and Hamilton, R. J. (eds.) (1989). *Rancidity in Foods*, 2nd edn, Applied Science, London.
2. Frankel, E. W. (1983). In *Recent Advances in the Chemistry of Meat*, (ed. A. J. Bailey), The Royal Society of Chemistry, London, p. 87.
3. Pearson, A. M., Gray, J. L., Wolzak, A. M. and Horenstein, M. A. (1983). *Food Technol. (Champaign)*, **37**, 121.
4. Proceedings of a Symposium Oxidative Rancidity of Meat and Meat Products (1987). *Food Sci. Technol. Today*, **1**, 151–173.
5. Lawrie, R. A. (1985). *Meat Science*, 4th edn, Pergamon, Oxford, p. 48.
6. Klettner, P.-G. and Baumgartner, P. A. (1981). *Food Technol. Aust.*, **32**, 380–4.
7. Evans, G. G. and Ranken, M. D. (1975). *J. Food Technol.*, **10**, 63–71.
8. Lawrie, R. A. (1985). *Meat Science*, 4th edn, Pergamon, Oxford, pp. 123–4.
9. Brooks, J. (1933). *J. Soc. Chem. Ind.*, **52**, 17T.
10. Ledward, D. A. (1984). In *Developments in Food Proteins—3*, (ed. B. J. F. Hudson), Elsevier, London.
11. Renerre, M. (1990). *Int. J. Food Sci. Technol.*, **25**, 613–630.
12. Lawrie, R. A. (1985). *Meat Science*, 4th edn, Pergamon, Oxford, p. 121.
13. Kendrick, J. and Watts, B. M. (1969). *Lipids*, **1**, 46.
14. Ledward, D. A. (1987). *Food Sci. Technol. Today*, **1**, 153.
15. McWeeney, D. J. (1968). *J. Food Technol.*, **3**, 15.
16. Lawrie, R. A. (1985). *Meat Science*, 4th edn, Pergamon, Oxford, p. 125.
17. Lawrie, R. A. (1985). *Meat Science*, 4th edn, Pergamon, Oxford, p. 156.
18. Lawrie, R. A. (1985). *Meat Science*, 4th edn, Pergamon, Oxford, p. 219.

19. Mottram, D. S., Patterson, R. L. S., Edwards, R. A. and Gough, T. A. (1977). *J. Sci. Food Agric.*, **28**, 1025.
20. Walters, C. L., Hart, R. J. and Perse, S. (1979). *Z. Lebensmitt. -Untersuch.*, **168**, 177.
21. Mottram, D. S. (1987). *Food Sci. Technol. Today*, **1**, 159.
22. Lea, C. H. (1962). In *Recent Advances in Food Research*, Vol. 1, (eds. J. Hawthorn and J. M. Leitch), Butterworths, London.
23. Mottram, D. S. and Edwards, R. A. (1982). *J. Sci. Food Agric.*, **34**, 517–522.
24. Pearson, A. M., Love, D. J. and Shorland, F. B. (1977). *Adv. Food Res.*, **23**, 1–74.
25. Johns, A. M., Birkinshaw, L. H. and Ledward, D. A. (1989). *Meat Sci.*, **25**, 209.
26. McLeod, J. A. (1982). *World's Poultry Sci. J.*, **38**, 194.
27. Crackel, R. L., Gray, J. I., Booren, A. M., Pearson, A. M and Buckley, D. J. (1988). *J. Food Sci.*, **53**, 656–657.

12 Legislation and labelling
C. HUMPHRIES

12.1 Introduction

There are numerous and detailed regulations, codes of practice and other legislative instruments worldwide controlling our food supply. The purpose of this chapter is not to provide a comprehensive guide but to give an insight into the most important principles on which they are based and, in particular, to discuss those areas of legislation which are pertinent to considerations of rancidity.

The chapter therefore starts with background information describing the purposes of prescribing legislation and introduces the Codex Alimentarius which provides internationally agreed guidelines and standards for food control, not forgetting reference to the European Community and its increasing importance in this area. Further sections deal with those areas of hygiene, processing and packaging, contamination, additives, particularly antioxidants and sequestrants, and labelling, including claims, which have a bearing on rancidity.

12.2 Principles of food legislation

Food law has three principal aims:

- the protection of the health of the consumer
- the prevention of fraud
- the promotion of fair trade

12.2.1 Background

The origins of food law as we know it today date back to the 19th century. In those days food safety was largely a matter of ensuring that food was free of acute toxicants, from substances such as lead and arsenic that were deliberately added to foods, often for the purposes of colouring products. Adulteration was also common. Substances were added to foodstuffs to make them go further; the addition of water to milk and other drinks, sawdust to bread or sand to pepper are examples. Other fraudulent practices developed as more sophisticated processing techniques allowed the separation of substances which may have an alternative use or command a higher price, such as the separation of cream from milk. Such practices led to the formulation in law of general offences, such as those of the Food Safety Act[1] of Great Britain which includes the following three offences;

(1) Any person who renders any food injurious to health by means of any of the following operations, namely:
    (a) adding any article or substance to the food;
    (b) using any article or substance as an ingredient in the preparation of the food;
    (c) abstracting any constituent from the food; and
    (d) subjecting the food to any other process or treatment,
with intent that it shall be sold for human consumption, shall be guilty of an offence (Section 7).

(2) Any person who:
    (a) sells for human consumption, or offers, exposes or advertises for sale for such consumption, or has in his possession for the purpose of such sale or of preparation for such sale; or
    (b) deposits with, or consigns to, any other person for the purpose of such sale or of preparation for such sale,
any food which fails to comply with food safety requirements shall be guilty of an offence (Section 8).

(3) Any person who sells to the purchaser's prejudice any food which is not of the nature or substance or quality demanded by the purchaser shall be guilty of an offence (Section 14).

12.2.2 Codex Alimentarius

Each country has developed its own rules to meet the needs of food control and safety. However, they are all based on the same general principles which are embodied in the Codex Alimentarius.[2] This is a collection of international food standards, adopted by the Codex Alimentarius Commission, an international body set up in 1962 jointly by the Food and Agriculture Organisations of the United Nations (FAO) and the World Health Organisation (WHO) to execute a programme aimed at protecting the health of consumers and facilitating international trade in foods. In like manner it prescribes that no food should be in international trade which:

    (a) has in or upon it any substance in an amount which renders it poisonous, harmful or otherwise injurious to health; or
    (b) consists in whole or in part of any filthy, putrid, rotten or decomposed or diseased substance or foreign matter, or is otherwise unfit for human consumption; or
    (c) is adulterated; or

(d) is labelled, or presented in a matter that is false, misleading or deceptive; or
(e) is sold, prepared, packaged, stored or transported for sale under unsanitary conditions.

These general principles are strengthened with detailed codes and guidelines for controls in specified areas. They are further elaborated with compositional standards for all the principal foods, whether processed, semi-processed or raw, for distribution to the consumer. They are widely accepted by countries and, in accordance with locally established legal and administrative procedures, form a basis for worldwide trade.

12.2.3 European Economic Community

As its name implies, the initial driving force in establishing the European Community (EC) was to facilitate trade. It was somewhat later that the desire to afford equal rights and protection to consumers throughout the member states was taken on board. Consequently, the initial thrust of the harmonisation programme for food was directed towards agreeing commodity standards and establishing a common list of permitted additives, both of which have proved notoriously difficult to agree. Only in recent years, with the launching of the single market initiatives in 1985[3], have we seen the development of so-called 'horizontal' or 'framework' directives which, like the general principles enunciated above, prescribe general parameters for food control leaving the more specific controls to the so-called 'vertical' directives which underpin them.

The importance of Community legislation must not be underestimated. Not only does it provide common food law for the 12 member states but it is increasingly recognised as a driving force by neighbouring countries, as witnessed by the signing of the treaty on the European Economic Area (EEA) in Spring 1992, extending the majority of the benefits of the EC to the 7 countries of European Free Trade Association (EFTA).

12.3 Hygiene

Sound handling and good hygienic practices throughout the food chain are essential if the optimum quality of raw materials and products is to be maintained. This is important in protection from and prevention of the development of rancidity.

12.3.1 Food handling

Handling foods and ingredients with care and using appropriate techniques will prevent spoilage, protect against contamination and minimise damage. All handling procedures should be such as to prevent raw materials, intermediates and finished products becoming contaminated. Primarily this protects against potential health hazards, but it also ensures the product remains in sound condition. Poor handling during harvest and production can result in damage to crops releasing enzymes, such as lipases and lipoxygenase, which promote rancidity (see chapter 8). Poor handling at any stage may trigger free radical formation resulting in the insidious development of rancidity regardless of other precautions taken at other stages in the food chain to prevent it. The minimisation of exposure to light or properly controlled low temperature storage are two examples of handling procedures that may be important in controlling rancidity.

Precautions should also be taken to protect against contamination. This is not just a question of avoiding incorporation of foreign matter but includes the avoidance of contamination with pests or by chemical, physical or microbiological contaminants. Chemical contamination can be particularly damaging if it is from traces of material such as copper, which promote free radical formation. Particular avoidance of contact with such materials during production and processing will be important in procedural considerations in handling materials prone to autoxidation. Contamination with organic material and micro-organisms, whilst undesirable from the potential health risks they present, should equally be avoided from the point of view of their potential to promote lipolytic rancidity.

12.3.2 Cleaning and disinfection

Regular cleaning to remove food residues and dirt, which may act as a source of contamination of establishments, equipment, (including utensils) and vehicles is critical in maintaining good hygiene. It may be followed or associated with disinfection. Any detergents and disinfectants used must be suitable for the purpose intended, since they may pose a risk of further contamination. Whilst the choice of these materials is usually left to the processor they can be prescribed in legislation or recommended in codes of good practice. Particular care must be exercised in their choice, especially when handling fatty materials which are prone to pick up taints and odours. For this reason phenolic compounds should be avoided.

12.3.3 Hygiene regulation

Regulation can either be product-specific or of a general nature. Its main purpose is to ensure the hygienic quality of the product. In general, regulations are provided to ensure:

(a) the cleanliness of the premises, vehicles and equipment used for the purposes of a food business;
(b) the hygienic handling of food;
(c) the cleanliness of persons engaged in the handling of food and of their clothing, and the action to be taken when they suffer from or are the carriers of certain infections likely to cause food poisoning;
(d) the construction of premises, vehicles and ships used for the purposes of a food business; their repair and maintenance;
(e) The provision of a clean and wholesome water supply and washing facilities;
(f) the proper disposal of waste material;
(g) the temperatures at which certain foods are to be kept.

General hygiene regulations based on these principles are found in the UK.[4-6] Internationally there is the Codex Alimentarius Code of Practice, 'General Principles of Food Hygiene'.[7] Until recently the EC has been preoccupied with developing a series of product-specific vertical directives for products of animal origin which include detailed provisions in relation to hygiene and health protection. A proposal for a general hygiene directive,[8] based on the above principles, and providing for the development of industry codes of practice, developed with reference to the provisions of the Codex Alimentarius code of practice, is close to being adopted.

12.3.4 Quality systems

Handling procedures, particularly those that are related to problems such as those relevant to minimising development of rancidity, are usually not matters for regulation but are left to the discretion of the entrepreneur. In recent years the adoption of quality systems, in particular ISO 9000, or its European (EN 29 000) and British (BS 5750) equivalents, and more recently total quality management systems, have been adopted by the food industry. These systems serve to ensure consistency of product and production.

12.3.5 HACCP

More important in maintaining hygiene and product quality is the adoption and implementation by the industry of Hazard Analysis Critical Control Points (HACCP). Recognition of the importance of this system is provided in the proposed EC general hygiene directive[8] which will, when adopted, give it force of law. Chapter 9 deals with HACCP in detail.

12.4 Irradiation

This processing technique, which can be used to disinfect food, inhibit sprouting, pasteurise or sterilise it, is singled out because it can lead to the formation of free radicals. Therefore the possibility of this being a source of rancidity must not be discounted where foods or ingredients have been irradiated. The technique has evoked much controversy and is tightly controlled. Its use is only permitted in certain countries, is limited to named foodstuffs and to maximum permitted doses. For this reason foodstuffs susceptible to oxidative rancidity are unlikely to be included on permitted lists.

12.5 Packaging

The choice of packaging material can also be important in preventing and controlling rancidity. Packaging provides a barrier which can control the atmosphere surrounding a product either by preventing or controlling the exchange of gases with the outside environment. Since oxygen is critical for autoxidation its elimination from stored products is a feature in extending the shelf life of vulnerable products. Ignoring other product requirements, this may mean selecting a material that prevents or controls the ingress of atmospheric oxygen such as would be used for vacuum packaging or regular packaging, or one which retained a modified or controlled atmosphere such as might be attained by nitrogen flushing or combined atmospheres of nitrogen and carbon dioxide. Packaging can also be used to protect against light.

Food may become contaminated by the constituents of packaging materials. General regulation seeks to prevent this by requiring that materials and articles intended to come into contact with food must not, under foreseeable conditions of use, transfer their constituents to food in quantities which could endanger human health or bring about an unacceptable change in the organoleptic properties of the food. This has been embraced in a regulation which applies throughout the European Community.[9,10]

The concern with potential health hazards from the transfer of constituents of packaging material has led to the further refinement of packaging legislation.[11] This will not only prescribe what substances can be used in packaging but also provides for maximum levels of migration of both specific constituents and a general, or global, migration limit for all migrating substances. The advantages of this for the control of rancidity will be a greater knowledge of the potential migration of constituents from different types of packaging where these constituents are known to promote rancidity.

12.6 Contaminants

Regulation of heavy metals has largely been concerned with health protection and has concentrated on setting maximum permitted levels for known toxicants such as lead[12] and arsenic[13] and more recently prescribing a maximum level for tin.[14] Purity criteria for additives also feature these metals but may also extend to pro-oxidant metals such as copper. New provisions provided for in the Food Safety Act[1] and currently being addressed in Brussels[15] will provide a wider remit to control contaminants in general. It would seem unlikely that these would impact directly on the control of rancidity unless the components concerned also posed a health hazard.

12.7 Additives

Food additive is defined by the Codex Alimentarius[2] as

> 'any substance not normally consumed as a food by itself and not normally used as a typical ingredient of the food, whether or not it has nutritive value, the intentional addition of which to food for a technological (including organoleptic) purpose in the manufacture, processing, preparation, treatment, packing, packaging, transport or holding of such food results, or may be reasonably expected to result, (directly or indirectly) in it or its by-products becoming a component of or otherwise affecting the characteristics of such foods'.

This same definition has been adopted by the EC.[16]

Usually additive controls are based on the positive list principle; that is, no additive other than those on a list of permitted additives may be used in or on food for human consumption. They may or may not be restricted to use in particular products and accompanied by maximum permitted levels of use. Once an additive is accepted onto a permitted list it is usual for it to be accompanied by specified purity criteria.

There are three fundamental criteria that must be met before additives can be approved for food use:

1. They present no hazard to the health of the consumer at the level of use proposed, so far as can be judged on the scientific evidence available.
2. They meet a technological need the purpose of which cannot be achieved by other means which are economically and technologically practicable.
3. They do not mislead the consumer.

12.7.1 Safety assessment

The generally accepted means of assessing the safety of additives, is by setting acceptable daily intakes (ADIs) and comparing these with predicted patterns of intake at normal/predicted levels of additive use. Maximum permitted levels on a product category basis can then be prescribed, taking account also of sub-groups of the population which may be more likely to be exposed, for example children or persons with special dietary needs such as diabetics. Where the toxicity of an additive is very low and the maximum possible use at normal usage levels does not present a hazard to health, an ADI is not specified. Such additives are usually permitted without restriction. In the EC the principle of 'quantum satis' has been adopted for these additives, meaning that no maximum level is specified but that they should be used according to good manufacturing practice and at a level no higher than is necessary to achieve the intended purpose, also providing they do not mislead the consumer.

The fact that an ADI can be established for a compound does not necessarily mean that the compound will be permitted. It remains for regulatory authorities to decide whether intakes can be controlled to ensure that levels set are not exceeded and whether additives should be permitted.

The effect of different dietary patterns and culturally divergent attitudes to food are at the root of current difficulties in harmonising additive legislation in the EC. Current regulations, which include specific provisions for antioxidants, include derogations to take account of member states' peculiarities. Eliminating these derogations is fundamental in ensuring the free circulation of goods throughout the Community, and compromises are having to be made to ensure that these new rules are not prejudicial to the health of the consumer or compromise the wide culinary diversity within the EC.

12.7.2 Relevant categories of additives

For the purposes of legislation and labelling, additives are assigned to categories according to their function in food systems. The category of most importance in considerations of rancidity is, of course, the antioxidants and certain preservatives which act as antioxidants. Packaging gases and sequestrants also have a role to play.

12.7.2.1 Antioxidants. A number of substances are permitted for use as antioxidants. For simplicity they can be divided into two categories depending on whether or not they can be found in natural products.

For the so-called 'natural' antioxidants, the tocopherols (tocopherol-rich extract, α-tocopherol, γ-tocopherol and δ-tocopherol) and the ascorbates (ascorbic acid, sodium ascorbate, calcium ascorbate, ascorbyl-palmitate and ascorbyl-stearate) an ADI is not specified and they are, for the most part, permitted unrestricted, *'quantum satis'* in the EC.

The use of most other antioxidants is restricted to named products or product categories and at specified maximum levels. All three parameters will vary depending on the country concerned. Tables 12.1 and 12.2 show the latest EC proposals for sorbates, benzoates and hydroxybenzoate (Table 12.1) and sulphur dioxide and sulphites (Table 12.2).

It is generally accepted that foods for babies, infants and young children must be controlled more rigorously because of the particular vulnerability of the consumer and their restricted dietary pattern. Table 12.3 illustrates the limited use of antioxidants currently included in EC proposals.[17]

12.7.2.2 Sequestrants. These are substances which form chemical complexes with metallic ions and as such have an indirect effect on development of rancidity. It is not usual for permitted lists to detail sequestrants as such. These are usually contained in lists such as those for 'miscellaneous additives' or 'other permitted additives'.

The substances concerned are principally acids and salts of weak acids; citrates, tartrates, lactates and malates and complexing agents such as EDTA. Legislation in the UK[18] has not so far restricted the use of these to particular commodities or maximum levels. The exception is calcium disodium ethylene diamine tetra-acetate (calcium disodium EDTA) which is restricted to use in canned fish, primarily to prevent the formation of a glass-like substance, 'struvite'. This will largely continue to be the case if

Table 12.1 EC proposed conditions for permitting sorbates, benzoates and hydroxybenzoates in foodstuffs[17]

(a) Definitions

EEC No	Name	Abbreviations
E 200	Sorbic acid	
E 202	Potassium sorbate	Sa
E 203	Calcium sorbate	
E 210	Benzoic acid	
E 211	Sodium benzoate	
E 212	Potassium benzoate	Ba[a]
E 213	Calcium benzoate	
E 214	Ethyl *p*-hydroxybenzoate	
E 215	Sodium ethyl *p*-hydroxybenzoate	
E 216	Propyl *p*-hydroxybenzoate	PHB
E 217	Sodium propyl *p*-hydroxybenzoate	
E 218	Methyl *p*-hydroxybenzoate	
E 219	Sodium methyl *p*-hydroxybenzoate	

Table 12.1 (*cont.*)

(*b*) *Conditions*

Foodstuffs	Maximum level (in mg kg^{-1} or mg l^{-1} as appropriate)					
	Sa	Ba	PHB	Sa + Ba	Sa + PHB	Sa + Ba + PHB
Dairy-based flavoured drinks	–	200	–	300 (of which 200 Ba maximum)	–	–
Wine-based flavoured drinks	100	–	–	–	–	–
Water-based or fruit-juice-based drinks	300[b]	150	–	250 Sa + 150 Ba[b]	–	–
Liquid tea concentrates and liquid fruit and herbal infusions concentrations	–	–	–	600	–	–
Alcohol-free wine	–	–	–	300	–	–
Grape juice, unfermented, for sacramental use	–	–	–	2000	–	–
Wines	200	–	–	–	–	–
Fermented fruit juices	500[b]	200	–	200 Sa + 200 Ba[b]	–	–
Mead	200	–	–	–	–	–
Spirituous beverages with not more than 15% alcohol content	200	200	–	400	–	–
Fillings of ravioli and similar products	1000	–	–	–	–	–
Jams, jellies, marmalades and similar products, energy-reduced or sugar-free and other fruit-based spreads	–	500	–	1000	–	–
Candied fruit	–	–	–	1000	–	–
Dried fruit	1000	–	–	–	–	–
Fruit-based desserts	–	–	–	1000	–	–
Fruit and vegetable preparations and fruit syrups	1000	–	–	–	–	–
Vegetables in vinegar, brine or oil	–	–	–	2000	–	–
Potato dough	2000	–	–	–	–	–
Meat products, cooked or cured or dried, surface treatment only (gelatin coatings included)	–	–	–	–	–	*Quantum satis*
Semi-preserved fish products	–	–	–	2000	–	–
Fish-roe products	–	–	–	2000	–	–
Shrimps, cooked	–	–	–	2000	–	–
Crangon crangon and *Crangon vulgaris* (Brown shrimp), cooked	–	–	–	2000	–	–
Milk, heat-treated fermented	–	–	–	300	–	–
Milk, renneted	–	–	–	1000	–	–
Cheese, prepacked sliced	1000	–	–	–	–	–
Unripened cheese	1000	–	–	–	–	–

Table 12.1 (*cont.*)

Foodstuffs	Maximum level (in mg kg^{-1} or mg l^{-1} as appropriate)					
	Sa	Ba	PHB	Sa + Ba	Sa + PHB	Sa + Ba + PHB
Processed cheese	2000	–	–	–	–	–
Dairy-based desserts	–	–	–	300	–	–
Liquid egg (white, yolk or whole egg)	–	–	–	5000	–	–
Pre-packed sliced bread and rye-bread	2000	–	–	–	–	–
Pre-baked bakery wares intended for retail sale	2000	–	–	–	–	–
Fine bakery wares with a water activity of > 0.65	2000	–	–	–	–	–
Cereal- or potato-based snacks and coated nuts	–	–	–	–	1000 (of which 300 PHB maximum)	–
Cake mixes	–	–	–	–	2000 (of which 300 PHB maximum)	–
Batters	2000	–	–	–	–	–
Sugar, nut or fat-based confectionery	–	–	–	–	–	1500 (of which 300 PHB maximum)
Cocoa-based confectionery (excluding chocolate)	–	–	–	1500	–	–
Chewing gum	–	–	–	1500	–	–
Sugar toppings (syrups for pancakes etc)	–	–	–	1500	–	–
Fat emulsions	–	–	–	2000	–	–
Emulsified sauces	2000	–	–	–	–	–
Non-emulsified sauces	–	–	–	1000	–	–
Salads	–	–	–	1500	–	–
Mustard	–	–	–	1000	–	–
Seasonings, condiments and mixed spices	–	–	–	1000	–	–
Soups and broths	–	–	–	500	–	–
Dietary food supplements	–	–	–	–	–	2000
Dietetic foods intended for special medical purposes—Dietetic formula for weight control intended to replace total daily food intake or an individual meal	–	–	–	–	1500	–

[a]Benzoic acid may be present in certain fermented products resulting from fermentation processes following good manufacturing practice.
[b]Either use Sa or a combination of Sa + Ba.
Notes
(1) The levels of all substances mentioned above are calculated as the free acid. (2) The abbreviations used in the table mean the following: Sa + Ba: Sa and Ba used singly or in combination; Sa + PHB: Sa and PHB used singly or in combination; Sa + Ba + PHB: Sa, Ba and PHB used singly or in combination. (3) The maximum levels of use indicated refer to foodstuffs ready for consumption prepared following manufacturers' instructions.

Table 12.2 Proposed conditions for permitting sulphur dioxide and sulphites in foodstuffs[17]

(a) Definitions

EEC No	Name
E 220	Sulphur dioxide
E 221	Sodium sulphite
E 222	Sodium hydrogen sulphite
E 223	Sodium metabisulphite
E 224	Potassium metabisulphite
E 226	Calcium sulphite
E 227	Calcium hydrogen sulphite
E 228	Potassium hydrogen sulphite

(b) Conditions

Foodstuffs	Maximum level expressed as SO_2 ($mg\,kg^{-1}$ or $mg\,l^{-1}$ as appropriate)
'Burger meat' with a minimum vegetable and/or cereal content of 4%	450
'Breakfast sausages' with a minimum cereal content of 6%	450
'Longaniza fresca' and 'butifarra fresca'	450
Dried salted fish of Gadidae spp.	200
Crustaceans and cephalopods	
fresh and frozen	150 ⎤ in edible
cooked	50 ⎦ parts
Fine bakery wares	50
Starches (excluding starches for weaning foods, follow-on formula)	50
Starches and modified starches for weaning foods, follow-on formula and infant formula	10
Sago	30
Pearled barley	30
Dehydrated granulated potatoes	400
Cereal and potato based snacks	50
Peeled potatoes	50
Processed potatoes (including frozen potatoes)	100
Potato dough	100
White vegetables, dried	400
White vegetables, processed (including frozen white vegetables)	50
White cardamom, cumin and caraway seed	500
Dried ginger	150
Dried tomatoes	200
Horseradish pulp	800
Onion, garlic and shallot pulp	300
Onions in vinegar	300
Other vegetables and fruits in vinegar, oil or brine	100
Processed mushrooms (including frozen mushrooms)	50
Dried fruits	
apricots, peaches, grapes, prunes and figs	2000
bananas	1000
apples and pears	600
other (including shell nuts)	500
Dried coconut	50
Peeled fruit and citrus peel	100
Candied fruit and candied citrus peel	100

Table 12.2 (*cont.*)

Foodstuffs	Maximum level expressed as SO_2 ($mg\,kg^{-1}$ or $mg\,l^{-1}$ as appropriate)
Jam, jelly and marmalade as meant in Directive 79/693/EEC[a]	50
Jam, jelly and marmalade made with sulphited fruit	100
Other fruit based sandwich spreads	50
Fruit based pie fillings	100
Citrus juice based seasonings	200
Concentrated grape juice for home wine making	2000
Fruit mustard, fruit curd and fruit chutney	100
Jellying fruit extract, liquid pectin for sale to the final consumer	800
Canned and bottled fruit (with heart cherries, mixtures with white heart cherries, rehydrated dried fruit and lychees and sliced lemon)	100
Sugars as meant in Directive 73/437/EEC[b]	15
except glucose syrup, whether or not dehydrated:	20
Other sugars	40
Sugar toppings (syrups for pancakes etc.)	40
Concentrates of citrus fruit juice, apple juice and pineapple juice	350
Orange, grapefruit, apple and pineapple juice	50
Lime and lemon juice	350
Concentrates based on fruit juice and containing not < 0.5% barley on a ready to drink basis	350
Other concentrates based on fruit juice or comminuted fruit	250
Non-alcoholic fruit juice based drinks	20 (carry-over from concentrates only)
Water based flavoured drinks containing glucose syrup as the main carbohydrate	50
Grape juice, unfermented, for sacramental use	70
Grape juice	15
Glucose-based confectionery	50
Beer, including low and alcohol free beer	20
Cask-conditioned beer	50
Wines	as in Regulations (EEC) No 822/87[c], 4252/88, 2332/92 and 1873/84
Alcohol-free wine	200
Cider, perry, fruit wine, sparkling fruit wine	200
Mead	200
Fermentation vinegar	170
Mustard, excluding Dijon mustard	250
Dijon mustard	500
Gelatin	50
Vegetable and cereal proteins	200
Vegetable and cereal protein-based meat, fish and crustacean analogues	200

Notes
(1) Maximum levels are expressed as SO_2 in $mg\,kg^{-1}$ or $mg\,l^{-1}$ as appropriate and relate to the total quantity, available from all sources. (2) An SO_2 content of not more than $10\,mg\,kg^{-1}$ or $10\,mg\,l^{-1}$ is not considered to be present.
[a] OJ No L 205, 13.8.1979, p. 5. [b] OJ No L 356, 27.12.1973, p. 71. [c] OJ No L 84, 27.3.1987, p. 1.

Table 12.3 EC proposed conditions for permitting antioxidants in foods for babies and infants[17]

	Antioxidant	Foodstuff	Maximum level
Infant formula and Follow-on formula	Tocopherol-rich extract α-tocopherol γ-tocopherol δ-tocopherol	All	10 mg l^{-1} singly or in combination
Weaning foods	L-Ascorbic acid Sodium L-ascorbate Calcium L-ascorbate	Fruit and vegetable-based drinks, juices and baby foods. Fat-containing cereal-based foods including biscuits and rusks	Singly or in combination expressed as ascorbic acid 0.3 g kg^{-1} 0.2 g kg^{-1}
	L-Ascorbyl palmitate Tocopherol-rich extract α-tocopherol γ-tocopherol δ-tocopherol	Fat-containing cereals, biscuits, rusks and baby foods	0.1 g kg^{-1} singly or in combination

current EC proposals[17] are adopted with extended controls for calcium disodium EDTA (Table 12.4).

The use of such substances in foods for babies, infants and young children is usually restricted to citric acid, with no maximum level prescribed.

12.7.2.3 Packaging gases. Argon, helium, nitrogen, nitrous oxide, hydrogen and oxygen can all be used as packaging gases and are included as such in EC proposals.[17] There are no restrictions, with the exception of nitrous oxide, which is not allowed in infant formula, follow-on formula and weaning foods. Carbon dioxide is also generally permitted.

Table 12.4 EC proposed conditions for permitting calcium disodium EDTA[17]

Foodstuff	Maximum level
Emulsified sauces	75 mg kg^{-1}
Canned and bottled pulses, legumes, mushrooms and artichokes	250 mg kg^{-1}
Canned and bottled crustaceans and molluscs	75 mg kg^{-1}
Canned and bottled fish	75 mg kg^{-1}
Minarine	100 mg kg^{-1}
Frozen and deep-frozen crustaceans	75 mg kg^{-1}

12.8 Labelling and presentation

Consumer protection is at the basis of food law. In previous sections we have seen, running in parallel with the need to ensure safety, the requirement not to mislead the consumer. This, and the need accurately to inform the purchaser about the product, are the principal reasons for controlling labelling and presentation. They have a secondary role in ensuring fair competition which is particularly important in the case of imports. Labelling rules must allow competing products to be distinguished.

12.8.1 General labelling requirement

The following list, taken from the EC Labelling Directive,[19] details the main parameters with which a product must be labelled:

- the name of the food,
- a list of ingredients,
- an indication of minimum durability,
- any special storage instructions or conditions of use,
- the name or business name and an address or registered office of the manufacturer or packer, or of a seller established within the European Community,
- the particulars of the place of origin if failure to give such particulars might mislead a purchaser to a material degree as to the true origin of the food, and
- instructions for use if it would be difficult to make appropriate use of the food in the absence of such instructions.

In addition most foods are required to be marked with a weight or volume and products on sale must display a selling price. Less stringent provisions apply where foods are sold loose, the name, certain additives and the price being the prime requirements.

12.8.2 Ingredients, durability and storage

These three parameters are of most interest with regard to rancidity and, to some extent, are interlinked.

Ingredient lists on foods must generally list, in descending order by weight, all of the ingredients, including additives, used in their preparation. Additives must be described by function or category name and by specific name (or number, where applicable). Thus all the antioxidants should be grouped together and included at the appropriate point in the list as; 'Antioxidants (potassium sorbate and sodium benzoate)'. Where an additive serves more than one function it is left to the labeller to decide upon the principal function and assign the additive accordingly. Therefore, although they have antioxidant functions, certain additives such as the sulphites, ascorbates and citrates are often not described as such in ingredient lists.

As a rule, additives contained legally in ingredients of foodstuffs which play no part in the finished foodstuff are not required to be labelled. An exception to this rule, which sets an important precedent, is the requirement under the EC 'jams' directive[20] which requires the labelling of sulphur dioxide when present in the final product at a level of 30 mg kg^{-1} or more.

So far certain functions, such as sequestrants, have not been required as category names. The new EC labelling proposals, when first published, included a requirement for the category names 'sequestrant' and 'packaging gas' to be included in the ingredients. Both have now been removed; sequestrant because it is meaningless to consumers and therefore not felt to be helpful or informative, packaging gas because these are not incorporated in the foodstuff itself. Sequestrants will continue to be required to be named in the ingredient list without a category name. Any declaration of packaging gas will be voluntary and will probably take the form of a claim, for example, 'packed in nitrogen for extra freshness'.

Minimum durability describes the shelf life of the product and is defined as the date up to and including which a food can reasonably be expected to retain its specific properties if properly stored. Therefore if the date stated is dependent upon particular storage requirements, such as chilled storage or keeping the product in an airtight container, this advice must also be included. Clearly the formulation, particularly the use of additives such as antioxidants, packaging gases and the packaging material itself have a direct influence on the date stated. Any alterations should lead to a reappraisal of the declaration.

12.8.3 *Claims*

A claim is any representation which states, suggests or implies that a food has particular characteristics relating to its origin, nutritional properties, nature, production, processing or composition or any other quality. This definition, taken from Codex Alimentarius guidelines[21] has also been adopted by the EC.[22]

Claims are used to promote and market foods. Their use in labelling and advertising is becoming particularly widespread especially in relation to quality, health and nutrition, which is fuelled by an increasing interest by consumers in diet and health.

The regulation of claims is necessary to prevent the consumer being misled, to protect against adverse health implications and to promote fair trading. Their proliferation has led many European countries to legislate independently in this area. As a consequence, EC proposals are being formulated.[22] Controls have been adopted in certain non-Community countries, particularly the United States where very detailed regulation on foodstuffs labelling[23] has recently been adopted.

Claims should only be made where they can be substantiated. It is generally accepted that food claims will not be made in relation to prevention, alleviation, treatment or cure of diseases, claims which imply that a balanced diet cannot supply all nutrients or claims that a food provides an adequate supply of all essential nutrients. Where claims are permitted, particularly in the case of nutrients, certain quantitative requirements may be specified. This is to ensure that the nutrient for which a claim is being made is present at a level which has physiological significance.

12.8.3.1 Polyunsaturates and cholesterol. Of particular interest in relation to rancidity are claims concerning fats, especially polyunsaturated fats, sometimes specifically highlighting the absence of cholesterol. Physiological criteria are satisfied by specifying the minimum fat content and the minimum proportions of polyunsaturates and maximum proportions of saturates which must be present.

Current thinking in Europe is based on the opinion that food cholesterol has very little, if any, influence on blood cholesterol which is the important risk factor in coronary heart disease. It is the level of saturated fat in the diet which is the risk factor. It is therefore likely that cholesterol claims will not be included in the EC directive. The situation in the US is different; cholesterol claims are common and are controlled by stringent provisions.

12.8.3.2 Vitamins. Other claims of relevance are those concerning vitamins, since additions of physiologically significant levels of ascorbates and tocopherols can qualify for vitamin C and E claims, respectively. Such levels are well in excess of those needed to control rancidity.

Vitamin claims are usually based on meeting a prescribed percentage of the recommended daily amount (RDA) and are often linked to a quantified serving. Like ADIs, RDAs are specified by officials on the basis of available information and vary from country to country, as do the prescribed percentage levels which qualify for 'high' and 'increased' claims. As far as EC proposals[22] are concerned, these levels look like being set at 30% and 15% of the RDA, respectively compared with current UK levels of 50% and a sixth.[24] Europe-wide RDAs are still under discussion. Levels have been specified in the Nutrition Labelling Directive (Table 12.5). Latest information suggests that the matter is being referred to Codex. In the meantime, the above values will be implemented for the purposes of nutrition labelling and claims.

Table 12.5 Vitamins and minerals which may be declared and their recommended daily allowances (RDAs)

Vitamin	RDA	Vitamin	RDA
Vitamin A	800 μg	Vitamin B12	1 μg
Vitamin D	5 μg	Biotin	0.15 mg
Vitamin E	10 mg	Pantothenic acid	6 mg
Vitamin C	60 mg	Calcium	800 mg
Thiamin	1.4 mg	Phosphorus	800 mg
Riboflavin	1.6 mg	Iron	14 mg
Niacin	18 mg	Magnesium	300 mg
Vitamin B6	2 mg	Zinc	15 mg
Folacin	200 μg	Iodine	150 μg

As a rule, 15% of the recommended allowance specified in the Annex supplied by 100 g or 100 ml or per package if the package contains only a single portion should be taken into consideration in deciding what constitutes a significant amount.

Whether it will, in future, be possible to make overt health claims about the role of different fats and vitamins in heart disease and cancer remains to be seen.

References

1. *Food Safety Act* (1990) (C 16).
2. *Codex Alimentarius*, Vol. 1 (1992). Food and Agriculture Organisation of the United Nations and World Health Organisation, Rome.
3. *Completing the Internal Market*, White Paper from the Commision to the European Council, COM (85) 310, 14.6.85.
4. *Food Hygiene (General) Regulations* 1970/1172 as amended by SIs 1982/1727, 1985/67, 1990/1431, 1990/2468 and 1991/1343.
5. *Food Hygiene (Markets, Stalls and Delivery Vehicles) Regulations* 1966/791 as amended by SIs 1966/1487, 1982/1727, 1985/67, 1990/1431, 1990/2486 and 1991/1343.
6. *Food Hygiene (Docks, Carriers etc) Regulations* 1960/1602 as amended by SIs 1962/1287, 1982/1727, 1985/67, 1990/1431 and 1990/2486.
7. *Recommended International Code of Practice General Principles of Food Hygiene*, CAC/RCP 1-1969, Rev. 2 (1985).
8. Amended proposal for a *Council Directive* on the hygiene of foodstuffs (92/C 347/11) (OJ No C 347; 31.12.92).
9. *Council Directive*, 89/109/EEC (OJ No L 40; 11.02.89).
10. *Commission Directive* 90/128/EEC (OJ No L 349/ 13.12.90).
11. *Commission Directive* 92/39/EEC (OJ No L 168; 23.6.92).
12. *Lead in Food Regulations* 1979/1254 as amended by SIs 1982/1727, 1985/67, 1985/912, 1990/2486 and 1991/1476.
13. *Arsenic in Food Regulations* 1959/831 as amended by SIs 1960/2261, 1962/1287, 1963/1435, 1972/1391, 1973/1052, 1973/1340, 1975/1486, 1982/1727, 1984/1304, 1985/67, 1990/2486 and 1991/1476.
14. *Tin in Food Regulations* 1992/496.
15. *Council Regulation* (EEC) 315/93 (OJ No.L37; 13.2.93).
16. *Council Directive* 89/107/EEC (OJ No L 40; 11.02.89).
17. EU Draft Common Position on Food Additives other than Colours and Sweeteners (1994), document 4552/94.

18. *Antioxidants in Food Regulations* 1978/105 as amended by SIs 1980/1831, 1983/1211, 1984/1304 and 1991/2540.
19. *Council Directive* 79/112/EEC (OJ No L 33; 08.02.79), as amended by Council Directives; 85/7 (OJ No L 2; 03.01.85), 86/197 (OJ No L 144; 29.05.86), 89/395 (OJ No L 186; 30.06.89), 90/496 (OJ No L 276; 6.10.90), 91/72 (OJ No L 42; 15.02.91).
20. *Council Directive* 79/693/EEC (OJ No L 205; 13.08.79, as amended by Council Directives; 80/1276 (OJ No L 37; 31.12.80), 88/593 (OJ No L 318; 25.11.88).
21. *Codex General Guidelines on Claims*, CAC/GL 1-1979 (Rev 1 – 1991).
22. Proposal for a *Council Directive* on the use of Claims concerning Foodstuffs (Doc. SPA/62/ORIG-Fr/Rev 2).
23. *Federal Register*, Vol. 58(3), (1993). January 6.
24. *The Food Labelling Regulations* 1984/1305 as amended by SIs 1985/71, 1985/67, 1987/1986, 1988/2112, 1989/768, 1989/2321, 1990/2486, 1990/2488, 1990/2489, 1991/2489 and 1991/1476.

13 Rancidity in creams and desserts
A. C. DAVIES and M. LIVERMORE

13.1 Introduction

Creams and desserts, in the context of this chapter, comprise oil-in-water emulsions which are either consumed liquid or, more usually, after aeration to form a whipped structure. In general, they contain a minor proportion of fat compared with other ingredients. The main sub-group are creams (sometimes referred to as 'non-dairy') used as alternatives to dairy cream, which is itself dealt with in a separate chapter. Biscuit creams are also not dealt with here, being of a different structure; they are covered in chapter 9.

Both short- and long-life products are included in this categorisation. Those with short shelf lives (measured in days) are pasteurised, whereas longer storage stablility may be achieved by:

1. Thermal sterilisation, either in-pack or in-line followed by aseptic packing.
2. Freezing.
3. Control of water activity via formulation; in this case products are normally plastic rather than liquid.
4. Drying.

13.2 Liquid creams

13.2.1 Short life

Pasteurised non-dairy creams are now not common, having been largely replaced by long-life, sterilised versions. However, consideration of this category will illustrate principles common to all liquid creams.

In general, these products consist of about 25% fat emulsified in a continuous aqueous phase containing milk solids, sugars, colours, flavours and stabilisers. Simple products may contain nothing more than colours, flavours and stabilisers in the water phase, but all such creams also contain emulsifiers to maintain a stable emulsion and to control whippability. Typical examples of formulations for such creams are shown in Table 13.1.

Products of this type are used as alternatives to dairy cream and, in some countries, dairy cream extenders. They are lower in cost, but also have certain functional advantages, being higher and more consistent in overrun and being less susceptible to collapse or serum weeping. Such products, when only pasteurised, have a chill shelf life measured in days, so the chemical processes of fat oxidation and rancidity development do not normally have a chance to give rise to any flavour deterioration.

Such creams have traditionally also been made from hardened lauric fats in order to give good whipping properties. The content of unsaturated (and, in particular, polyunsaturated) fatty acids is very low, and hence the stability to oxidation is high. Palm kernel oil hardened to a slip point of 38 degrees and having an iodine value of 3–6 is commonly used.

Although stable against oxidation, the oils used in these products must still be of very high quality, since the final emulsions are made unpalatable by even low levels of off-flavours. In particular, lauric fats are prone to deterioration by hydrolytic mechanisms to give rise to soapy flavours.

Non-dairy creams do deteriorate by attack on the fatty phase, however, but by microbiological action rather than by autoxidation. Fat splitting by lipase activity can result in off-tastes, particularly soapy tastes, when lauric fats have been used. Means of avoiding this are by thorough pasteurisation of creams at source, hygienic control of processing and packaging and by use of high quality raw materials. The use of stable vegetable fats to replace dairy fats in some cream-type products is described by Lautsen.[1]

Table 13.1 Typical formulations for long-life, sterilised creams

Ingredients	Content (%)
Recipe A	
Methyl cellulose	0.6
Sodium carboxymethyl cellulose	0.3
Hardened palm kernel oil	30.0
Colour/flavour	as desired
Water	to 100
Recipe B	
Hardened coconut oil	22.0
Emulsifiers:	
Sodium stearoyl lactylate	0.3
Polysorbate 60	0.1
Lactylated monoglyceride	0.6
Sucrose	6.0
Sodium caseinate	2.0
Salt	0.1
Sodium carboxymethyl cellulose	0.4
Water	68.4

Rancidity in both dairy and non-dairy systems is related to short-chain fatty acids (C_{12} and below) as produced by lipolysis (see Scanlan et al.[2]). A commony-used analytical measure of free fatty acids is the acid degree value (ADV) as described by Richardson.[3] However, doubts have been raised as to whether this has a good correlation with perceived rancidity (for example, Duncan et al.[4]) because of the different contribution to this flavour of individual acids, and thus the only fully reliable assessment method at present is taste panelling.

Although the nature and intensity of the rancid flavour will depend upon the type of fat, the presence of lipolytic bacteria is clearly something to be avoided in all cases. Lipase activity can be monitored and used as a check on product quality and process efficiency. Various methods for measuring lipase activity are available, but the method particularly recommended for creams using lauric fats is shown below. As suitable prepared tributyrin agar may not be commercially available, it is usually necessary to prepare the required medium to the formulation and method shown below. The agar is stabilised by the addition of a small amount of polyvinyl alcohol:

Medium:
- Distilled water, 1000 ml
- Lab Lemco powder, 5 g
- Tryptone, 10 g
- NaCl, 5 g
- Yeast extract, 3 g
- Glycerol tributyrate, 10 ml
- Polyvinyl alcohol, 1.5 g
- Oxoid agar bacteriological No. 3, 15 g

The tryptone, sodium chloride, Lemco powder and yeast extract are dissolved, with stirring, in 750 ml of distilled water. The glycerol tributyrate and polyvinyl alcohol are emulsified in the remaining 250 ml of water by mixing for 10 min with a high speed mixer at 25 000 rpm. The first solution is added to the emulsion and the whole mixture stirred for a further 5 min. Finally, the Oxoid agar No. 3 is added and steamed to dissolve. The pH is adjusted to 7.0 and the medium dispensed into 150 ml bottles and autoclaved at 121 °C for 20 min. The final pH should be 7.0.

Method:
1. Pour the necessary number of plates of the medium to be used.
2. Dry the plates at 45 °C for 10 min.
3. Use a sterile 1 ml pipette and dispense 0.1 ml of the inoculum on the surface of the plate.
4. Take a glass spreader from its pot of alcohol and flame it. Cool the

spreader before use by placing on the surface of the agar. Use this to spread the inoculum evenly over the agar surface.
5. Incubate at 30 °C for 2 days.

Positive presence of lipolytics is indicated by a transparent halo surrounding each colony. Results should mostly be completely negative, but a standard of less than 10 ml^{-1} is usually found acceptable.

To monitor lipase activity in milk or whey powders which may be used as raw materials, a fluorimetric method has been developed by the National Institute for Research in Dairying (NIRD), Reading.[5]

Although, as mentioned previously, lipase activity may only be deleterious to flavour when low-chain length fatty acids are released, it is claimed that further microbiological attack on fatty acids by β-oxidation (see chapter 1) and decarboxylation can produce methyl ketones with very high flavour potency and extremely low threshold values. In practice, however, actual instances of this happening seem to be very rare. An interesting paper by Alford et al.[6] is quoted for further reference.

13.2.2 Long life

Liquid creams of the types described can be produced in long-life form, e.g. six months shelf life at ambient temperature, by sterile processing and packaging. Over such a period, problems with true oxidative deterioration, as opposed to those of lipolytic activity, can certainly occur. The nature of the processing and packaging are, however, very favourable to product stability. In the Alpura sterilisation process, for instance, flash evaporative cooling of the sterilised product will remove oxygen to a very low level, whilst the packs themselves, e.g. Tetrapak, are impermeable to air and to light. If good quality, well-refined and fresh oil is not used for production, however, reversion flavours, sometimes described as 'cardboard-like', may develop. This can be completely avoided, however, if the tight specification and quality control standard described previously is rigidly maintained.

Many creams are effectively sterile versions of the short-life products described above, and have similar compositions. Modifications may be made to improve long-term emulsion stability, but fat blends remain the same and there are no other factors likely to increase the chance of rancidity development. End of shelf life is normally determined by physical performance, for example reduction of overrun, and rancidity development is very rare. Indeed, problems of oil flavour stability can be eliminated by good refining and storage of the oil prior to use; any flavour problems normally become evident shortly after production.

More recently, high quality non-dairy creams have become available, often with similar lauric-based fat blends, but using relatively high levels of dairy ingredients, particularly buttermilk, as part of the composition. Buttermilk (the liquid obtained when cream is churned to form butter) is relatively rich in phospholipids from the milk fat globule membrane (MFGM). It has been shown, e.g. by Chen,[7] that the MFGM can catalyse the oxidation of milk fat at elevated temperature. Because of the storage of non-dairy creams at ambient or lower temperatures, this is not in itself a problem. However, poor handling of the buttermilk can result in rancidity development in the butterfat with consequent off-flavour in the cream.

The practice with high quality creams is to store them chilled to maintain emulsion stability through their shelf life. This will, of course, reduce the risk of rancidity even further.

The other means for converting liquid non-dairy creams into long-life form is by freezing. This is carried out by some commercial operators but, of course, requires the formulation of a product with satisfactory freeze/ thaw properties, usually by the inclusion of solutes such as sugar. At the temperatures involved, problems with rancidity should not be encountered.

All of these liquid creams are of short life after use on cakes, desserts, etc. Only 1–3 days at chill temperature is involved, during which time development of fat off-tastes should not be encountered with normal precautions.

13.3 Solid creams

By solid creams we mean plastic emulsions of fat (25–40%) in fairly saturated sugar solutions. They may be flavoured and coloured and are either ready to whip directly or after addition of extra water, i.e. as concentrates. They are used in a range of applications from heavy cake fillings to light dessert toppings. Functionally similar products may be produced in a water-in-oil form by emulsifying sugar syrup in the fat phase, in this case, usually greater than 50% fat. With flavour and colour added, such products are sometimes referred to as 'butter creams'.

All of these product types are pasteurised during production and have low water activity in the aqueous phase. They are vulnerable, therefore, to osmophilic yeasts and moulds. Although lauric fats may be used in these products, their steep melting characteristics do not lend themselves to the achievement of plastic texture over a reasonable temperature range, so that blends of liquid and hydrogenated vegetable oils, possibly including lauric fats, are more often used, not too dissimilar from margarine blends.

The life of such products is usually only a few weeks at ambient temperature, so that normal precautions in selection of good quality oils and fats will avoid problems with rancidity. First evidence of deterioration is generally mould growth, even at the low water activity of aqueous phase normally used. Use of sorbate at a suitable pH can be helpful, but good hygiene at the user's premises, e.g. the bakery or kitchen, is very important, as cases are known of product rejection because of off-tastes caused by rapid lipolytic activity, traced back to poor cleaning of equipment.

13.4 Powders

So called 'powdered fats' are used in the formulation of many powdered toppings, creams, desserts, etc. Powdered fats are interesting as they represent the type of product where oxidative stability of the fat can be the critical factor in determining product shelf life. In general, powdered fats made by spray-drying of oil-in-water emulsions tend to have good stability, as the fat is protected by a complete outer coating of dry aqueous phase—milk powder, caseinate, maltodextrin, etc.—and an ambient shelf life of several months should readily be achieved.

A number of factors are important in this achievement, however, particularly the use of suitable packaging and storage conditions. Clearly, fats with good stability should be used. Hydrogenated lauric fats are extensively included, but these may be replaced by higher fatty acid-containing, steep-hardened alternatives such as hydrogenated palm oil, if soapy rancidity is especially feared because of any particular adverse circumstances, e.g. climate. Phosphates are often used in the formulation of these products, not primarily for, but conferring nevertheless, a useful sequestering effect on trace metals which may be present. A helpful factor may be the known inhibitory effect of casein on lipid oxidation. The protection of dried whole milk powder from oxidised flavour, described as 'like cardboard', has been reported by Chan.[8] The use of Maillard reaction products and of butylated hydroxyanisole (BHA)/butylated hydroxytoluene (BHT) gave favourable results, but the most significant improvement appears to have been by storage under nitrogen.

Precautions to avoid the onset of oxidative rancidity include the following product and packaging aspects.

13.4.1 Product

1. Good quality, stable oils must be used, as fresh as possible after refining.

2. Hydrogenated oils are desirable, with minimum content of polyunsaturated fatty acids.
3. Stability can be enhaced by the use of permitted antioxidants, especially where natural tocopherols may be absent or at a low level. Note that the antioxidants should be added as soon as possible after refining if maximum benefit is to be achieved; they will not remove oxidation products which have already been formed.
4. The quality of encapsulation (which is related to the exclusion of air) is very important. In powdered, oil-based flavours, for example, it has been demonstrated that gum arabic can be more effective than a maltodextrin, and the protection given by the encapsulation medium can be equal to the use of vacuum or inert gas packaging. In combination with a suitable antioxidant, this should result in achievement of excellent shelf life.

13.4.2 Packaging

1. Use of appropriate packaging, for the storage conditions and shelf life required, is of vital importance.
2. The use of vacuum or inert gas headspace, in packs which are impermeable to both light and oxygen, is ideal.
3. Instructions for cool storage are highly desirable.

The kinetics of fat oxidation in whole milk powder during storage is being investigated at the Swedish Food Institute at Gothenberg.[9] Whilst results so far reported showing the effects of antioxidant and nitrogen headspace are too complex to provide clear guidance for storage of fat-containing powders, the work will be of considerable interest to operators in this field.

13.5 Conclusions

As opposed to many other fat-containing systems, the main source of rancid taste development is often microbiological rather than oxidative. Hygienic manufacture, packaging, storage and use will avoid problems.

For long-life products in liquid form, process and packaging usually provide good insurance against oxidation.

For powders, the spray-drying process itself confers good protection to the fat. If lipolytic activity is feared, then non-lauric fats such as hardened palm oil may be used.

Antioxidants and sequestering agents are not usually necessary, but can provide extra protection, as also can sorbate mould inhibitor.

References

1. Lautsen, K. (1986). Vegetable fats in the dairy industry, *Dairy Ind. Int.*, **51** (2), 12–3.
2. Scanlan, R. A., Sather, L. A. and Day, E. A. (1965). Contribution of free fatty acids to the flavour of rancid milk. *J. Dairy Sci.*, **58**, 1582.
3. Richardson, G. H. (1985). *Standard Methods for the Examination of Dairy Products*, 15th edn, American Public Health Association, Washington DC.
4. Duncan, S. E., Christen, G. L. and Penfield, M. P. (1991). Rancid flavour of milk: relationship of acid degree value, free fatty acids and sensory perception. *J. Food Sci.*, **56** (2), 394.
5. Stead, D. (1984). A fluorimetric method for determination of *Pseudomonas fluorescens* AR11 lipase in skim milk powder, whey powder and whey protein concentrate, *J. Dairy Res.*, **51** (4), 623–8.
6. Alford, J. A., Smith, J. L. and Lilly, H. D. (1971). Relationship of microbial activity to changes in lipids of foods, *J. Appl. Bacteriol.*, **34** (1), 133–46.
7. Chen, Z. Y. and Nawar, W. W. (1991). Role of milk fat globule membrane in autoxidation of milk fat. *J. Food Sci.*, **56** (2), 398.
8. Chan, H. W. S. (1978). *Autoxidation of Unsaturated Lipids*, Academic Press, London, p. 215.
9. Hall, G., Andersson, J., Lingnert, H. and Olofsson, B. (1985). Flavour changes in whole milk powder during storage. II. The kinetics of the formation of volatile fat oxidation products and other volatile compounds, *J. Food Qual.*, **7** (3), 153–90.

14 The control of rancidity in confectionery products
F. B. PADLEY

14.1 Introduction

The consumer expects confectionery products to exhibit good flavour, texture and appearance. In order to deliver these expectations the confectionery manufacturer has to consider the impact of many factors on final product characteristics including type and quality of raw material, product fabrication, processing and storage. Depending on the product, its shelf life will be determined normally by texture and flavour stability. Appearance, notably gloss of chocolate, can also be a factor, especially when the product is subjected to extremes of temperature during storage.

Flavour is a prime characteristic used by the consumer to judge quality and acceptability. It is therefore essential for the food technologist and manufacturer to understand the factors influencing flavour stability and the causes of rancidity and loss of fresh flavour if they are to succeed consistently in producing products of defined quality and shelf life. As with many fabricated foods the causes of rancidity in confectionery products arise from two main sources.

The first is autoxidation of lipids in which oxygen becomes bound to a fatty acyl chain in the form of a hydroperoxide. This usually occurs under the influence of a catalyst such as a heavy metal, or a lipoxygenase present say in a nut or kernel. Degradation of hydroperoxides gives rise to short-chain fragments which in many cases can have exceedingly low flavour thresholds having unpleasant tastes.

The second form of rancidity is microbial in origin and can arise in at least two ways. For example either via lipolysis under the action of microbially generated lipases releasing short-chain fatty acids with a 'soapy' taste or by microbial conversion of acids to compounds having very low flavour thresholds. The causes of rancidity are therefore numerous and complex. This chapter is therefore divided into a number of sections describing the role of major ingredients and the relevance of autoxidation to fat deterioration. Specific fats of major interest to the confectioner are discussed individually. Separate sections are devoted to nuts and dairy ingredients which are major components of most high quality confections and finally a general section covers other aspects not

THE CONTROL OF RANCIDITY IN CONFECTIONERY PRODUCTS   231

covered in preceding sections. Microbial initiated rancidity is dealt with under those ingredient sections where it is most pertinent, notably lauric fats, nuts and dairy products.

14.2 The role of major ingredients

14.2.1 Polysaccharides

Although superficially these might be regarded as inert there are specific interactions with proteins which influence flavour perception. Most notable are the flavours which are developed as a consequence of the Maillard reaction. Although high temperatures are associated with this reaction there are instances where, over long periods at ambient temperatures, Maillard reaction products have been observed.

The Maillard reaction involves a condensation reaction between reducing sugars and proteins. The subsequent degradation and modification of Amadori compounds result in the formation of brown pigments and flavours ranging from burnt, hay-like to caramel. The reaction sequence is complex and dependent on temperature and pH[1] (see Figure 14.1). Flavours arising from the Maillard reaction are in most cases used in a positive way to manipulate the flavour of products. Those generated under long-term storage conditions will be received in a more negative way and perceived as a loss of fresh taste. They are in part, responsible for flavour deterioration in products stored for long periods. However this is almost certainly one of the lesser causes of off-flavour deterioration in confectionery products (see section 14.3 on role of water activity) unless it arises from poorly controlled storage and processing, thus introducing an undesirable flavour attribute to the product. Similarly caramelization of sugar alone is also a vehicle for flavour development, created by heating to high temperatures. Provided that the caramelization temperature is properly controlled and tailored to the product characteristics it is regarded as a positive contributor to flavour.

14.2.2 Protein

The two sources of protein in confectionery are from vegetables, e.g. nuts, and dairy products. Vegetable protein arises from cocoa beans and other vegetable seeds such as soya, hazelnut, peanut etc. Proteins by themselves are not normally associated with a source of off-flavour unless it arises from microbial degradation or high temperature hydrolysis. Proteolysis for example can lead to the production of low molecular weight peptides with low flavour thresholds. Bitter notes in milk have for example been attributed to degradation of proteins by bacteria/enzymes.[2]

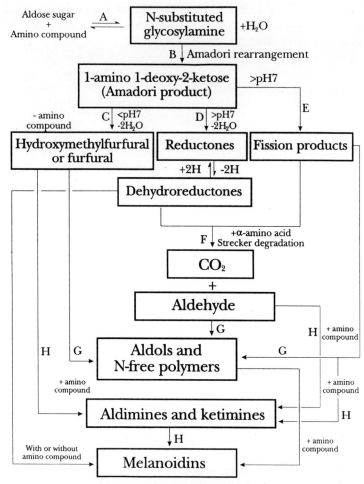

Figure 14.1 Non-enzymic browning.[1]

Proteins interact with sugars as discussed above under the Maillard reaction and this will be the main source of flavours arising from protein in confectionery.

14.2.3 Oils and fats

A wide variety of oils and fats are used in confectionery (Table 14.1). Although relatively saturated fats form the major proportion of fats used, e.g. cocoa butter, there will be occasions when more unsaturated fats such as sunflower are used. The main cause of flavour deterioration in fats is autoxidation and some key aspects are therefore covered here.

Table 14.1 Typical fatty acid composition of fats and oils

	6:0	8:0	10:0	12:0	14:0	16:0	16:1	18:0	18:1	18:2	18:3	20:0	20:1	22:0	22:1	24:0	Others
Almond						6.7	0.5	1.2	66.3	22.3	Tr	0.3					3
Brazil						14.1	0.3	8.6	29.0	46.6	0.1	Tr		Tr		Tr	1.1
Butter	2.2	1.2	2.8	3.0	10.2	25.0	2.6	12.1	27.1	2.4	2.0	1.0	0.5	0.2	Tr	Tr	8.9[a]
Cocoa Butter					0.1	26.0	0.3	34.3	34.8	3.0	0.2	0.1					
Coconut	1.3	12.2	8.0	48.8	14.8	6.9		2.0	4.5	1.4		0.1					
Corn				0.1	0.2	13.0		2.5	30.5	52.0	1.0	0.5	0.2	0.1		0.4	0.9
Cottonseed					0.8	27.3	0.8	2.0	18.3	50.5	Tr	0.3					
Groundnut						12.5	Tr	2.5	37.9	41.1	0.3	0.5	0.7	2.5	1.0	1.0	
Hazelnut						4.7	0.2	1.6	76.4	16.3	0.1	0.1	0.1				0.4
Illipe butter					0.2	17.5	0.2	46.0	35.0	1.0		0.1					
Olive (Turkey)						11.1	0.7	2.5	74.1	9.4	0.2	0.4	0.2	0.1		0.4	0.9
Palm (Nigeria)				0.2	1.0	45.0	0.1	4.6	37.7	10.6	0.2	0.3					0.3
Palm Kernel	1.0	3.0	4.0	49.0	16.0	8.0		2.5	14.0	2.5							
Pecan						5.7	0.1	2.2	66.9	22.1	1.1	0.2	0.4	Tr			1.3
Rape (low eruc)						4		2	56	26	10	Tr	2	Tr	Tr		
Sal						5.2		42.8	40.0	3.8	0.8	7.0					
Shea nut					Tr	3.3	Tr	44.3	45.6	5.5		1.3					
Soya						11.0	0.5	4.0	22.0	53.0	7.5	1.0	1.0	0.9		0.2	
Sunflower					0.1	5.5	0.1	4.7	19.5	68.5	0.1	0.3	0.1				
Walnut						3.1	Tr	2.6	29.1	58.3	4.9						2.0

[a]Includes butryic (4:0) and various branched-chain acids.

14.2.3.1 Autoxidation. The autoxidation of oils and fats and the subsequent products of hydroperoxide decomposition have been widely studied and reviewed.[3,4] It is therefore not the purpose here to discuss the area in detail, more to highlight some specific effects that occur in confectionery. It is widely accepted that free radical autoxidation readily occurs in the presence of heat, light, irradiation and initiators such as heavy metals. In an industrial situation, raw materials are unlikely to be subjected to light or irradiation. Problems are more likely to arise due to the combined effects of heat, oxygen, and/or the presence of heavy metals. For example the storage of bulk oils under inadequately controlled conditions, e.g. oils stored at say 60 °C or above in the presence of air and agitated either continuously or periodically will lead to appreciable oxidation and flavour deterioration. The extent to which this occurs will also depend on the level of antioxidants, natural or otherwise, in oil and the degree of unsaturation of the fatty acids. The relative rates of free radical initiated autoxidation for different acids are given in Table 14.2.

Autoxidation is catalysed by the presence of trace amounts of heavy metals and it is normal to specify maximum level of iron (≤ 0.1 ppm) and copper (≤ 0.01 ppm) in refined oils. However refined oil quality and flavour stability is not simply measured in terms of conventional parameters such as free fatty acid, peroxide value, heavy metals etc. Oil quality and its subsequent flavour stability will also depend on the presence and level of flavour precursors generated by possible oxidative action prior to oil extraction and refining, i.e. autoxidation occurring via the action of lipoxygenase in stored seeds.[5] To some extent this has been resolved by the use of both peroxide value (PV) and anisidine value (AV) in defining oil quality. The latter measurement provides an estimation of the presence of fatty aldehydes arising from peroxide decomposition. In the case of palm oil totox values are often quoted and are defined as totox = 2PV + AV. Typical specifications for partially and fully refined palm oils are given below (see Table 14.3).[6]

Oils and fats and confectionery components containing them therefore need to be stored under controlled conditions, ideally under nitrogen, in the dark, to ensure flavour stability is extended for as long a period as

Table 14.2 Relative rates of oxidation of fatty acid methyl esters

Acid	Relative oxidation rate	
	Autoxidation	Photo-oxygenation
Oleic *cis* $\Delta 9$	1	3×10^4
Linoleic *cis* $\Delta 9$ *cis* $\Delta 12$	27	4×10^4
Linoleic *cis* $\Delta 9$, *cis* $\Delta 12$, *cis* $\Delta 15$	77	7×10^4

Table 14.3 Requirements for partially and fully refined palm oils entering Europe[6]

Characteristic	Caustic neutralized, still to be dried bleached and deodorized	RBD[a] to be blended as such
FFA	0.1% max	0.05% max
Cu	< 0.05 mg kg^{-1}	< 0.05 mg kg^{-1}
Fe	< 0.1 mg kg^{-1}	< 0.1 mg kg^{-1}
Totox	10 max	10 max
Colour bleaching test (Lov. R 5¼″)	bleaching test 3R[a]	3
Moisture	< 0.7%	< 0.05%
Taste	Not specified	Neutral, no off-flavour

[a]Physically refined and deodorized. [b]Specific bleaching test; 150 °C, 60 inf. 5% earth.

possible. Conditions for oil storage have been described in detail by Patterson.[7]

It is widely known that the presence of natural antioxidants has a significant effect on oil stability. The vast majority of confectionery products contain relatively low levels of polyunsaturated acids (< 5%) and significant amounts of tocopherols (> 200 ppm) and are therefore observed to have good flavour and oxidative stability. In this respect butterfat is potentially prone to autoxidation because it contains only low levels of natural antioxidants. In most confectionery products however cocoa butter has a strong stabilising influence. Typical levels of tocopherols found in some oils and fats are given in Table 14.4. In certain cases additional protection can be obtained by supplementing naturally occurring tocopherols with additional tocopherols or by adding synthetic

Table 14.4 Tocopherols in vegetable oils (mg kg^{-1}) (compiled from Carpenter[8] and Muller-Mulot[9])

Oil	α-T	β-T	γ-T	δ-T	α-T$_3$	γ-T$_3$
Cocoa butter	11	–	170	17	2	–
Coconut[a]	–	–	–	2–4	20	–
Corn[a]	134	18	412	39	–	–
Cottonseed	573	40	317	10	–	–
Groundnut	169	5	144	13	–	–
Palm[b]	279	–	61	–	274	3
Rape[a]	70	16	178	7	–	–
Safflower	477	–	44	10	–	–
Sesame[a]	12	6	244	32	–	–
Soybean[a]	116	34	737	275	–	–
Sunflower[a]	608	17	11	–	–	–
Wheat germ	1179	398	493	118	Tr	–

[a]Refined oils. [b]δ-T$_3$, 69 mg kg^{-1}.

antioxidants such as butylated hydroxy anisole (BHA) or tertiary butyl hydroquinone (TBHQ) provided legislation allows. Further protection is afforded by the use of metal sequestrants such as citric acid. The ability of lecithin to stabilise oils is explained in part by its ability to suppress the catalytic activity of copper and iron and also its synergistic effect with antioxidants.

It is useful at this point to comment on photo-oxidation although it is not normally associated as a problem with confectionery products because as a matter of common practice they are excluded from light.

14.2.3.2 Photo-oxidation. Photo-oxidation is clearly distinguished from the more conventional free radical autoxidation. Fats when exposed to UV light and in the presence of oxygen will undergo autocatalytic oxidation via attack by singlet oxygen. The reaction of singlet oxygen progresses at a rate approximately 1500 times that of normal oxygen.[3] Moreover the products of oxidation via singlet oxygen and the subsequent profile of the volatile flavour components arising from the hydroperoxide breakdown will also be different from those produced by normal free radical attack.

Photo-oxidation is quenched by a number of naturally occurring compounds, notably tocopherol and carotenoids. Under certain conditions, e.g. in the dark, carotenoids (>20 ppm) can however promote free radical autoxidation leading to a deterioration of flavour and odour. Other trace components such as sulphur-containing compounds, are also susceptible to photo-oxidation leading to the generation of rubbery off-flavours. Some specific oils and fats which are major components of confectionery products are considered later.

14.2.3.3 Estimating the flavour stability of oils and fats. It can take an extremely long time for most oils and fats to develop oxidative rancidity. For example the induction period at room temperature for even polyunsaturated oils such as groundnut oil can be many months. Confectionery manufacturers have therefore turned to using accelerated tests for measuring oxidative stability. There is no substantive evidence to show that accelerated tests such as the Rancimat test carried out at an elevated temperature can be used in any absolute way to predict flavour stability at ambient temperature. There is however a general consensus that these tests can be used to rank oils of the same type, e.g. all cocoa butters.[10] The relationship between the Rancimat induction period with temperatures (Figure 14.2) has led to a proposal that a 30 h induction period at 100 °C corresponds to a one year induction period at 20 °C.[10] However, one must then take note of the composition of the oil under test, in particular the presence of trace levels of highly unsaturated acid and

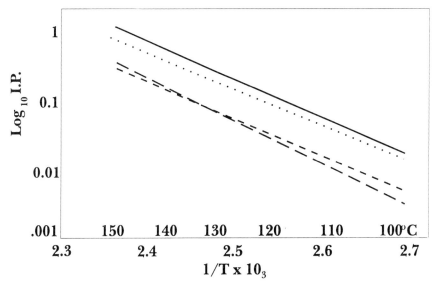

Figure 14.2 Arrhenius plots showing variation in induction period with temperature. ——, Brazilian cocoa butter; · · · · Non-deodorised cocoa butter; – – –, 65% beef tallow/35% palm oil; — — Palm/sea-based CBE.[10]

relevant practical data concerning the oils used in a given product. In the author's view all accelerated tests should be used with caution unless a clear correlation has been established, for example as Rossell notes; misleading results will be obtained if volatile antioxidants such as BHA or BHT are present.

14.3 The influence of water activity on flavour stability

The roles of the major ingredients have been discussed and so far only the chemical effects arising from autoxidation and chemical reaction between components have been considered. It is useful at this stage to consider the role of water and its effect on both autoxidation and on microbial activity. The role of water in food systems is complex and this short summary can hardly do it justice. However some of the more important factors can be emphasised.

The effect of water activity on food deterioration has been reviewed by Gould.[11] The relationship between water activity and food degradation pathways is summarised in Figure 14.3. This shows that although microbial growth is virtually zero at a water activity (a_w) below 0.6 some of the

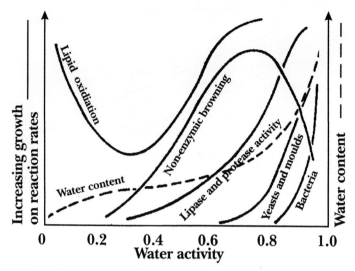

Figure 14.3 General water activity dependence of quality deterioration factors in foods. (Compiled by Gould[11] from refs. 35–37.)

other causes of flavour deterioration notably oxidation and non-enzymic browning continue. Indeed, under appropriate conditions as a_w is lowered oxidation can even accelerate. The effect of a_w on autoxidation[4] has been explained by proposing:

(i) At high a_w water interacts with metal catalysts reducing their activity.
(ii) Water interferes with the decomposition of hydroperoxides; at high a_w hydrogen bonding helps to stabilise the hydroperoxide.

This effect will not necessarily be observed in all food systems as the mechanism of autoxidation will differ to some extent depending on the type of catalysts involved. However it does emphasise the need to control the level of heavy metal autoxidation catalysts to a minimum in confectionery products.

The effect of water on the Maillard reaction has also been studied. Superficially the Maillard, or Browning reaction might be related to high temperature effects. In reality the Maillard reaction occurs even at ambient temperatures. In a model study using glucose and casein at 35 °C the Maillard reaction rate accelerated to a maximum value as water activity was reduced below 1.0, and then fell to zero at even lower water activities (Figure 14.4). Flavours generated as a consequence of the Maillard reaction will depend on the rate of degradation of the initially formed Amadori compounds.[1] The breakdown of Amadori compounds is

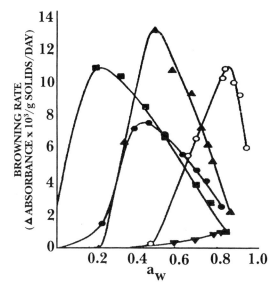

Figure 14.4 Browning rate versus water activity and moisture content for a casein–glucose model system containing 10% humectant. ○, control; ●, glycerol; ▲, 1,3-butylene glycol; ■, propylene glycol; ▼, sorbitol.[38]

related to temperature and pH and under the typical storage conditions used for confectionery this is likely to be a slow process. The area where problems might arise is at the production stage if partly formulated ingredients or recycled material are stored at elevated temperatures resulting in the development of uncharacteristic flavours.

To conclude this section on the role of water activity it is important to note its importance in relation to microbial growth. Microbial deterioration of confectionery can present significant problems if the appropriate precautions are not taken, e.g. ensuring proper hygiene. Having appropriate control procedures on raw materials and the combination of process design and product formulation can essentially eliminate microbial activity. It is in the area of product formulation that confectionery is afforded a significant measure of protection as a consequence of its low water activity. The effect of a_w on the growth of the major micro-organisms is given in Table 14.5. It is important to note that the *Eurotium* species for example grow, although slowly, at an a_w as low as 0.62, an observation discussed in more detail in sections 14.4.3 and 14.5 on lauric fats and nuts, respectively. The additional consequence of operating with products at low a_w is that not only is the growth of micro-organisms retarded but their thermal stability is increased and hence their viability in what might otherwise be regarded as conditions leading to sterilisation.[11]

Table 14.5 Low water activity limits for growth of micro-organisms, solute effects and equivalent osmotic pressures[11]

Water activity	Micro-organism	Predominant solute	Approximate pressure (MPa) equivalent osmotic
0.985	*Campylobacter jejuni*	NaCl	3
0.97	*Pseudomonas fluorescens*	Sucrose	6
	Pseudomonas aeruginosa	NaCl	
	Clostridium botulinum E	NaCl	
0.96	*Pseudomonas fluorescens*	NaCl	8
	Vibrio parahaemolyticus	Sucrose	
0.95	*Vibrio parahaemolyticus*	NaCl	10
	Clostridium botulinum A	NaCl	
	Clostridium perfringens	NaCl	
	Clostridium novyi	NaCl	
	Bacillus cereus	Glucose	
	Salmonella oranienberg	NaCl	
	Staphylococcus aureus	Erthyritol	
0.94	*Pseudomonas fluorescens*	Glycerol	
	Vibrio parahaemolyticus	Sucrose	
	Clostridium botulinum B	NaCl	
	Bacillus cereus	NaCl	
	Streptococcus faecalis	NaCl	
0.93	*Clostridium botulinum* A	Glycerol	
	Clostridium perfringens	Glycerol	
	Clostridium novyi	Glycerol	
	Staphylococcus aureus	Xylitol	
0.92	*Bacillus subtilis*	Glycerol	
	Bacillus cereus	Glycerol	
	Saccharomyces cerevisiae	NaCl	
0.90	*Bacillus subtilis*	NaCl	
	Saccharomyces aureus	Sucrose	
0.89	*Staphylococcus aureus*	Glycerol	
0.87	*Staphylococcus aureus*	Sucrose	
0.86	*Staphylococcus aureus*	NaCl	
		Sodium acetate	
0.80	*Saccharomyces* species	Dried foods	
	Aspergillus species		
	Pencillium species		
0.76	*Basipetospora halophila*	Glycerol	
0.75	*Halobacterium* species	NaCl	
	Wallemia sebi	Glycerol	
	Hansenula anomala	Glucose	
	Basipetospora halophila	Glucose and fructose	
0.73	*Eurotium* species	NaCl	
	Basopetospora halophila	NaCl	
0.70	*Eurotium* species	Glycerol	60
	Saccaromyces bisporus	Glucose	
0.66	*Eurotium* species	Glucose and fructose	

Table 14.5 (cont.)

Water activity	Micro-organism	Predominant solute	Approximate pressure (MPa) equivalent osmotic
0.66	*Eurotium* species	Glucose and fructose	
0.65	*Saccharomyces rouxii*	Glucose	
	Aspergillus echinulatus	Glucose	
	Torulopsis candida	Glucose	
0.62	*Eurotium* species	Dried foods	80
	Monacsus species		

14.4 Individual confectionery fats

14.4.1 Cocoa butter

Cocoa butter is the major fat component in chocolate and will be introduced by one or more of the following routes:

- as a component of cocoa mass
- as a component of cocoa powder
- as pressed cocoa butter
- as deodorised cocoa butter

Cocoa butter is generally recognised as a fat having high oxidative stability. This arises from both its low level of polyunsaturated acids and the presence of very effective natural antioxidants (Table 14.4) and metal chelators. Problems of flavour stability will probably arise for example if the following cocoa ingredients are used in confectionery:

- damaged beans containing fat with a high free fatty acid and peroxide value
- cocoa powder made from low quality beans and having a high microbial count
- pressed cocoa butter arising from poor quality beans therefore having a high FFA content and PV
- inadequately refined cocoa butter
- cocoa butter which has been stored in the molten state for extended periods etc.

However, cocoa butter is normally very resistant to off-flavour development because of its oxidative robustness and the normally high standard of quality control applied. Perhaps this is best exemplified by the long flavour stability associated with both chocolate and cocoa powder. In addition, cocoa lipids are subjected to oxidation under controlled conditions during roasting and conching. Conching of dark chocolate is normally carried out at 70 °C over a number of hours, if not days. These processes are clearly not sufficiently harsh to exhaust the protection afforded by natural antioxidants.

The role of accelerated tests in determining cocoa butter stability has been discussed above (Figure 14.2).

It is interesting to note that cocoa butter does not contain an exceptionally high level of tocopherol. Cocoa beans contains ca. 0.5% phospholipid of which 40% is phosphatidyl choline, 30% phosphatidyl inositol, 15% phosphatidyl ethanolamine plus other minor components. Pressed cocoa butters will also contain phospholipid (ca. 0.1%). Part of the oxidative protection arises almost certainly from the presence of polyphenols, related compounds and lecithin which can act as antioxidants and as metal sequesterants. In addition cocoa butter is held in the solid state both during most of its storage and in the finished product. The autoxidation of fats in the solid state will be retarded by virtue of the reduced access to oxygen. There will also be some enrichment of tocopherols by preferential partition into the small amount of liquid phase present in the fat, thus affording greater protection of the more unsaturated components.

14.4.2 Butterfat and dairy ingredients

Butterfat is a key ingredient of many confectionery items and a fat potentially prone to major rancidity problems. Again, butterfat is introduced in a diverse number of ways for example:

- as milk or cream followed by dehydration; e.g. crumb process
- as milk powder
- as condensed milk
- as dehydrated butter
- as butterfat, refined or non-refined
- as butterfat fractions etc.

Butterfat itself suffers from a severe deficiency of natural antioxidants (17 ppm α-tocopherol). Great care should therefore be taken to ensure that heavy metals such as copper and iron are essentially absent (< 0.01 ppm Cu, < 0.1 ppm Fe). Butterfat is clearly afforded protection in mixtures with cocoa butter by the presence of cocoa butter tocopherols. Ingredients and final products are nevertheless prone to off-flavour development in a number of ways.

It is one thing to control refined oil quality prior to product formulation and manufacture, but the consequence of product manufacture can create a situation in which autoxidation is facilitated. This is particularly true in the case of heavy metals which can easily find their way into a finished product by inadequate attention to the use of metals used in the fabrication of equipment. For example, butterfat, which is normally associated with the generation of soapy rancidity, is also prone to a 'fishy', metallic-type rancidity. This was characterised by Swoboda and Peers,[12-14] whose study highlights the nature of the problem food technologists encounter when manufacturing foods with an anticipated long shelf life. A fishy taint was detected in caramels whose formulation included a blend of refined vegetable oil and butterfat. The caramels contained 6 ppm of copper. The fishy taint only developed however if butterfat, copper and a significant level of tocopherol (1000 ppm) were all present.

The compounds responsible for the fishy taint were identified by isolating and separating the flavour volatiles generated in model experiments. The metallic odour is due to the vinyl ketones oct-1-en-3-one and octa-1-*cis* 5-dien-3-one. Oct-1-en-3-one is detectable at a 1 part in 10^{10} dilution and octa-1-*cis*-5-dien-3-one detectable at a dilution of 1 part in 10^{12} in aqueous solution. The fishy taint arises from the presence of the aldehyde deca-*trans*-2-*cis*-4-*cis*-7-trienal. The very low flavour threshold of these compounds is yet another example of the profound sensitivity of the human olfactory system and highlights the need to ensure heavy metal contamination is avoided.

One aspect which has not been fully explained to the author's knowledge is the role of copper in many traditional methods of producing boiled confectionery. Copper vessels have in the past been used by the industry with apparently acceptable results. One possible explanation is that high levels of copper are reported to have an antioxidant effect (e.g. autoxidation chain termination).

Regarding the introduction of butterfat via dairy products, a more complex situation is created and the conventional standards of hygiene must clearly be used in handling milk, cream or milk powders to minimise any occurrence of soapy rancidity which arises from microbial lipolysis (Table 14.6). In addition subsequent processing such as drying must be carried out under carefully controlled conditions to avoid caramelization, Maillard flavours and enhanced lactone formation.[17] The causes of undesirable off-flavours in dairy products have been reviewed.[18]

Milk and cream, as have been mentioned, are clearly prone to development of soapy rancidity which arises through lipolytic attack. Pasteurisa-

Table 14.6 Flavour characteristics and flavour threshold values of short-chain free fatty acids (Al-Shabibi et al.,[15] Kinsella[16])

Fatty acids	Flavour characteristic (in milk)	Flavour threshold (ppm)^2	
		Milk	Oil
Butyric	Butyric	25.0	0.6
Caproic	Cowy (goaty)	14.0	2.5
Caprylic	Cowy, goaty	–	350
Capric	Rancid, unclean, bitter, soapy	7.0	200
Lauric	Rancid, unclean, bitter, soapy	8.0	700

tion and careful hygiene is used to control this problem. As a general recommendation only the highest quality raw milk should be used in confectionery manufacture. Even seasonal variation in quality occurs as indicated by the level of free fatty acid (Table 14.7).[19] The development of bitter notes arises from microbial and proteolytic attack.[2] Microbial action is also responsible for a range of other flavour notes, e.g. fruity, green, musty, malty, arising in milk and dairy products.[18] Heat treatment of dairy products, e.g. pasteurisation, UHT treatment, can also generate a variety of off-flavours including 'cabbage' or 'sulphurous', cooked, scorched etc.[20,21]

In addition to milk and cream a wide variety of processed dairy products are used in confectionery manufacture including powders and liquid concentrates. Compositions of typical dairy components are given in Tables 14.8 and 14.9.

Table 14.7 Seasonal variation in FFA levels in milk supplies[19]

Months	FFA level (BDI) m mol/100 g of fat		Annual milk deliveries (%)
	Bulk tank milk	Can milk	
Jan/Feb	1.18	0.85	3
Mar/Apr	1.02	0.86	16
May/June	0.66	0.54	32
July/Aug	0.68	0.50	29
Sept/Oct	1.03	0.85	16
Nov/Dec	1.23	0.92	4

Table 14.8 Average composition (%) of milk-based products (powder)[17]

	Water	Protein	Fat	Lactose	Ash
Products containing total milk proteins					
Whole milk	2.0	25.4	27.5	38.2	5.9
Skimmed milk	3.0	35.8	0.7	51.6	7.9
Buttermilk	3.0	34.3	5.3	50.0	7.6
Coprecipitate	4.0	87–91	0.6–1.0	0.5	3.5–14.0
Caseinates	4.0	87–91	1.0–2.0	0.5	3.5–5.0
Whey protein products					
Sweet whey	4.0	12.0	1.3	73	7.9
Acid whey	4.0	12.0	0.8	69	11.5
Whey protein concentrates (UF)					
A	4.6	36.2	2.1	46.5	7.8
B	4.2	63.0	5.6	21.1	3.9
C	4.0	81.0	7.2	3.5	3.1
Whey protein isolate	5.0	92.0	0.3	0.3	2.8
Demineralized whey					
(90% demin.)	3.8	14.0	0.7	83	0.9
(70% demin.)	4.0	12.0	1.5	79	2.7

Table 14.9 Composition of condensed milks[39]

	Whole sweetened, (%)	Skimmed (non-fat sweetened), (%)	Unsweetened (evaporated), (%)
Fat	9.3	0.6	10.5
Sugar (sucrose)	41.0	43.0	–
Lactose	11.4	15.0	11.8
Protein	9.3	10.2	9.5
Ash	2.0	2.2	2.0
Water	27.0	29.0	66.2

Milk powder is notorious for the development of a cardboard-like off-flavour which arises via either a Maillard-type reaction (sugar–protein interaction) or via fat autoxidation, the latter route being supported by the observation that either the addition of antioxidants or storage under nitrogen delays the onset of this type of off-flavour.[23] Hall and Lingnert studied the change in flavour profile of dried whole milk when stored under a variety of conditions. The behaviour of products stored under nitrogen is given in Figure 14.5. Off-flavours can also develop in milk powders stored at low (-20 °C) temperatures. This has been explained on the basis of residual water freezing and leading to a concentration of heavy metals.

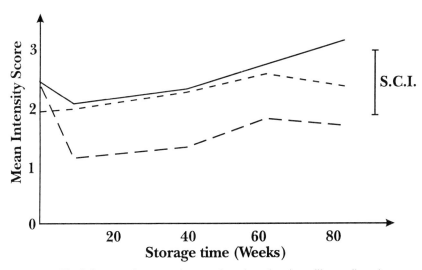

Figure 14.5 The influence of storage time on the odour descriptor-like cardboard. ——, no addition of antioxidant: air; — —, no addition of antioxidant: nitrogen; ---, 0.0025% BHA/BHT added: air. SCI, simultaneous confidence interval (5%).[23]

14.4.3 Lauric fats, soapy and ketonic rancidity

The lauric fats based either on palm kernel or coconut oils probably have the longest association with confectionery apart from cocoa butter or butterfat. The sharp melting stearin fraction from these fats, particularly palm kernel stearin, has a long history of application as a cocoa butter substitute in the manufacture of couvertures. The fact that many thousands of tonnes of lauric-based coatings are produced each year throughout the world allows us to put the off-flavour problems associated with these fats into perspective.

Lauric fats contain only low levels of unsaturated acid and are not normally associated with oxidative rancidity. However it should be borne in mind that these fats, in particular stearin fractions, contain only a low level of tocopherol and this should ideally be supplemented. Autoxidation will occur even with lauric fats if they are stored under adverse conditions, e.g. high temperature in air, for any length of time. Protection by the addition of tocopherol or another appropriate antioxidant is therefore recommended.

The main criticism of lauric fats is however their susceptibility to soapy or ketonic rancidity. Soapy rancidity arises in products which contain both an active lipase and moisture. The soapy taste is caused by the short-chain fatty acids released when the fat is partially hydrolysed. The flavour threshold of short chain acids is quite low (Table 14.6). Active lipases are present as a consequence of microbial contamination of one or more of the confectionery ingredients other than the lauric fat.

In virtually all cases soapy rancidity can therefore be controlled. However for those exceptions where potential contamination or moisture are to be encountered caution should be exercised.

The most common ingredient introducing lipases into lauric-based confectionery is poor quality cocoa powder; the effect can therefore be eliminated by adopting appropriate thermal treatment to inactivate both micro-organisms and lipases, using sterile cocoa powder and hygienic manufacturing processes. The generation of free fatty acid due to lipolytic activity is covered in more detail under the section describing nuts.

Ketonic rancidity has been rediscovered with recent work by Kinderlerer.[24,25] Most instances of rancidity in lauric products will arise from a combination of both ketonic and soapy (free fatty acid) rancidity. Some of the micro-organisms responsible for ketonic rancidity in coconut and desiccated coconut have been identified as xerophilic fungi, i.e. *Eurotium amstelodami*, *E. chevalieri*, *E. herbariorum* and *Pencillium citrinum*. These fungi degrade (Figure 14.6) the even-chain length fatty acids to a homologous series of aliphatic methyl ketones and secondary alcohols (C_5-C_{11}).[25,26] These off-flavours (Table 14.10) were developed in model systems stored at a water activity (a_w) of 0.76.

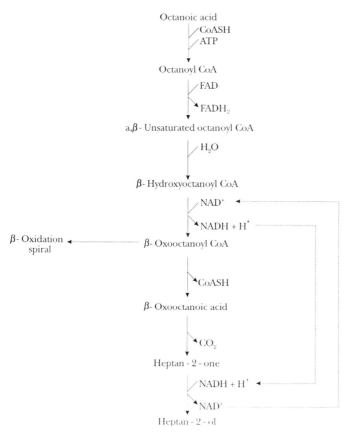

Figure 14.6 Modification of β-oxidation pathway of fatty acids to show the production of methyl ketones and secondary alcohols (conversion of octanoic acid to heptan-2-one and heptan-2-ol).

Table 14.10 Volatile off-flavour compounds in rancid coconut oil[26]

Off-flavour compounds	Consumer return (mg g^{-1} oil) s.d.	Rancid coconut (mg g^{-1} oil) s.d.	Flavour note	Intensity 1:4
Pentan-2-one	Trace	Trace	Pear drops	3
Pentan-2-ol	N.D.	0.58 ± 0.21	Ethereal	2
Hexan-2-one	9.36 ± 5.01	Trace	Ethereal	4
Hexan-2-ol	4.88 ± 5.45	N.D.	Turpentine	3
Heptan-2-one	10.72 ± 3.55	5.05 ± 0.68	Rancid almonds	4
Heptan-2-ol	9.29 ± 1.11	7.05 ± 0.18	Rancid coconut	4
Octan-2-one	2.57 ± 0.44	N.D.	Weakly ethereal	2
Nontan-2-one	19.63 ± 0.88	2.97 ± 0.07	Weakly turpentine	1
Nontan-2-ol	2.91 ± 0.88	0.35 ± 0.15	Musty, stale	2
Undec-2-one	6.77 ± 0.20	1.29 ± 0.12	Weakly turpentine	1
Total	66.13	17.29		

The solution is therefore clear. Ingredients and products must be essentially free of micro-organisms and free lipase. Products should be designed in such a way that microbial growth and lipase activity arising from contamination are inhibited by the appropriate control of water activity and where possible pH.

14.4.4 Confectionery fats other than CBE and lauric fats

There is a very wide range of confectionery fats available to the food manufacturer. Many of these fats are based on partially hydrogenated oils, so-called hard butters. In order to create fats with a sharp melting point the hydrogenation is carried out under conditions which maximise *trans* content. In most cases this also results in the near elimination of polyunsaturated acids. Such products, provided they contain antioxidant, e.g. tocopherol (100–200 ppm minimum) will exhibit good flavour stability.

14.5 Nuts and seeds

The combination of chocolate with nuts has a wide consumer appeal and this is reflected in the variety of nuts and seeds used in the confectionery industry (Table 14.11). A detailed compilation of nuts and their properties has been written.[28,29]

The main causes of rancidity in nuts arise via two avenues, microbial attack and autoxidation either under the influence of lipoxygenase or metal catalysis. The flavour and stability of nuts is therefore very dependent on the ways they are handled and stored. Mechanical damage creates open surfaces of ruptured cells which provide free oil which is open to both microbial/lipolytic attack and oxidative degradation especially under conditions of high moisture content.

Microbial attack of even whole nuts occurs however under conditions of high water activity. Nuts should therefore be stored at a_w of ≤ 0.6 in order to minimise or halt rancidity arising from microbial and lipolytic degradation. These aspects are discussed in the later part of this section.

Autoxidation of oil via the action of lipoxygenase which is present in most nuts has been highlighted in seeds of soya, groundnut and sunflower.[5] The generation of hydroperoxides either by the action of lipoxygenase or via metal catalysed autoxidation is one of the first steps towards off-flavour development in stored nuts in the absence of other causative factors.

For example soya bean is the most widely studied seed in the context of lipoxygenase. More limited studies have been carried out on seeds and nuts more appropriate to confectionery. It should therefore not be assumed that the oil in an intact nut is fully protected against oxidation and that it is only when the oil is extracted does oxidation begin.

With respect to metal catalysis it is interesting to note that in most nuts there is a significant level of copper and iron (ca. 10 ppm) arising in part from their occurrence as functional components in enzymes; lipoxygenase for example contains iron as Fe^{2+} or Fe^{3+}.

Numerous studies have been carried out on nuts in order to define conditions which maintain good flavour and texture. Provided water activity is controlled, the main cause of rancidity is autoxidation. It is therefore not surprising that rancidity is delayed by storing nuts at low temperature and under nitrogen. This is demonstrated in the work of Fourie and Basson[30] (Figure 14.7) who observed both the high oxidative stability of almonds compared to pecan nuts and the increase in peroxide and loss of unsaturation during storage at 30 °C at 55% humidity. The reasons for differences in flavour stability shown by different types of nuts are undoubtedly complex. Both nut oils have a similar fatty acid composition, iodine values, and contain ca. 22% linoleic acid with only trace amounts of more highly unsaturated acid (Table 14.11). However on a relatively simplistic basis the stability of almonds compared to pecan nuts can, at least in part, be related to the level of tocopherols (vitamin E as reported by McCance and Widdowson).[31] The almond contains six times the level of tocopherol expressed as vitamin E. The naturally occurring level of copper and iron is also at the ppm level and therefore of potential significance, especially if the nuts are crushed.

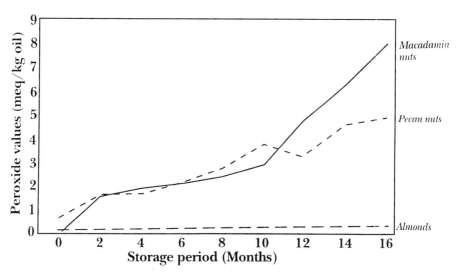

Figure 14.7 Mean values (10 determinations) for changes in peroxide values of different nuts during storage, nuts stored at 30 °C, 55% RH.[30]

Table 14.11 Fatty acid composition of tree nut oil[a,b,27]

Common name	Botanical name	Oil	Fatty acid (wt %)										
			16:0	16:1	18:0	18:1	18:2	18:3	20:0	20:1	22:0	22:1	24:0	others
Almond	*Prunus amygdalus*	52.7	6.7	0.5	1.2	66.3	22.3	tr	tr	tr	tr	c	—	3.0
Butternut	*Juglans cinerea*	60.0	1.6	tr	0.8	19.0	61.9	—	—	—	—	—	—	0.7
Brazil	*Berthrolletia myrtaceae*	68.7	14.1	0.3	8.6	29.0	46.6	tr	0.3	tr	tr	—	—	1.1
Cashew	*Anarcardium occidentale*	47.7	10.2	0.4	8.5	60.9	18.3	—	0.8	tr	tr	—	—	0.9
Chestnut	*Castanea mollisima*	2.5	14.6	0.7	1.1	54.0	24.9	2.7	—	1.0	0.2	—	—	0.8
Filbert	*Corylus avellana*	62.3	4.7	0.2	1.6	76.4	16.3	0.1	0.1	0.2	—	—	—	0.4
Hickory	*Caryaovata ovata*	70.4	8.8	0.5	2.3	52.0	33.5	1.7	0.2	tr	—	—	—	1.0
Macadamia	*Macadamia tetraphylla*[c]	73.2	8.3	21.8	2.1	56.4	2.8	—	2.4	3.1	0.8	0.3	0.5	1.3
Pecan	*Carya illioensis*	70.3	5.7	0.1	2.2	66.9	22.1	1.1	0.2	0.4	—	—	—	1.3
Pignolia	*Pinus sp.*	48.1	5.8	0.2	3.8	38.1	46.4	0.8	0.7	0.9	—	—	—	3.3
Pistachio	*Pistachio vera*	57.0	8.6	0.7	2.3	68.8	17.8	0.3	0.3	0.6	tr	—	—	0.6
Walnut, black	*Juglans nigra*	59.0	3.1	tr	2.6	29.1	58.3	4.9	tr	tr	tr	—	tr	2.0
Walnut, English	*Juglans regia*	67.4	7.3	0.2	2.3	19.1	57.4	13.1	tr	—	—	—	—	0.6

[a]Beuchat, L. R. and Worthington, R. E. (1978). *J. Food Technol.*, **13**, 355. [b]All fats contained tr (trace) of 17:0. [c]0.1% 17:1.

The flavour stability of hazelnuts under different storage conditions has also been reported.[32] Some typical results are shown in Figures 14.8 and 14.9. Off-flavours also arise as a consequence of microbial attack. In the case of coconut the micro-organisms responsible and the off-flavours

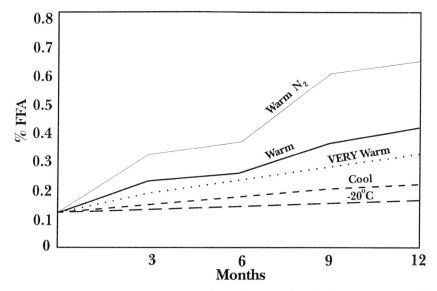

Figure 14.8 Changes of FFA-values of hazelnuts, Roman variety, during storage. --- cool, dry (3–6 °C; 50–60% RH); ———, warm, dry (18–25 °C; 50–55% RH); ·······, very warm, dry (35 °C; 30–40% RH); ———, warm, nitrogen (18–25 °C; N$_2$, 60–70% RH); – –, –20 °C; control.[32]

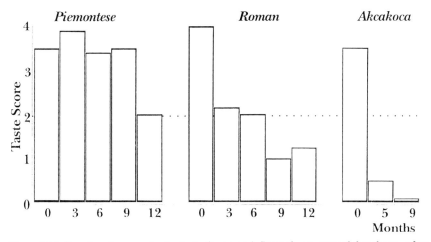

Figure 14.9 Results of organoleptic tests (taste and flavour) on ground hazelnuts, after different periods of storage under exposed conditions (18–25 °C, 50–65% RH). Taste score: 0, completely spoilt; 1, detrimentally affected, not acceptable; 2, detrimentally affected, still acceptable; 3, satisfactory; 4, very good.[32]

themselves have been characterised. This can be regarded as a general problem for all oil seeds, especially stored under conditions of high water activity (> 0.6). The micro-organisms and flavours responsible for rancidity in coconut have already been discussed under lauric fats. Rancidity in hazelnuts can be due to either microbial or oxidative attack. A study of Turkish hazelnuts revealed contamination by 28 fungal species, fungal spoilage being largely due to the *Penicillium* and *Erotium* species.[33]

However, in a separate study rancidity was seen to be primarily due to autoxidation.[34] The rancidity in nuts is therefore delayed by addressing the problems arising from both autoxidation and microbial attack. Nuts should therefore be stored under conditions of low water activity (< 0.6), at a low temperature and ideally under nitrogen if lengthy storage of many months is anticipated. For example large volumes of nuts are stored under refrigerated conditions to preserve quality and flavour.

Accelerated autoxidation tests have been explored by a number of workers in order to provide more rapid methods of determining shelf life. Different varieties of peanut were compared[22] by observing the rate of oxygen consumed by the nuts held at 99 °C. A good correlation with normal long-term storage behaviour at ambient temperature was observed. Similar conclusions have been reported for Turkish hazelnuts.[35] In this latter case the Rancimat apparatus was held at 120 °C. Typical results are given in Table 14.12. Kerassunda nuts were noted for their lower induction periods and shorter shelf life. Unfortunately in most of the work reported to date there was no attempt to relate induction period to other characteristics of the nut, e.g. oil content, fatty acid composition, free fatty acid, initial peroxide value and tocopherols etc.

Roasted nuts, not surprisingly, have a lower oxidative stability than unroasted nuts and products containing roasted, chopped nuts can become rancid fairly quickly especially if the nuts are chopped before roasting.[35]

Table 14.12 Induction periods of hazelnuts grown in Turkey[35]

6 month-old hazelnuts	Induction period (h)
FAQ Ordu 1980 crop	14
FAQ Ordu 1980 crop (shrivelled)	11.5
FAQ Akcakoca 1980 crop	16
FAQ Akcakoca 1980 crop (shrivelled)	12.5
13–15 mm Ordu 1980 crop	16
11–13 Samson 1980 crop (shrivelled)	11.5
9–12 mm Kerassunda 1980 crop	12
9–12 mm Kerassunda 1980 crop (shrivelled)	8.75

5 g of ground nuts placed in Rancimat flask which is immersed in an oil bath at 120 °C, with 10 l h^{-1} air passing through the contents of the flask. FAQ, fair average quality.

Table 14.13 Ingredient composition of some confectionery products[35]

Ingredients	Toffee	Nougat	Praline	Biscuit cream filling	Wafer	Cracknell	Truffle	Milk chocolate	Non-chocolate coating
Sugar	20.0	33.0	43.0	60.0	–	60.0	32.0	42.0	42.0
Fat	8.0	5.0	10.0	34.0	2.0	–	13.0	21.0	32.0
FCMP	–	10.0	–	–	–	–	10.0	25.5	–
SMP	–	–	6.0	–	–	–	3.0	–	17.7
SCMS	24.0	–	–	–	–	–	–	–	–
Glucose solids	30.0	42.0	–	–	–	25.0	18.0	–	–
Invert solids	8.0	4.0	–	–	–	–	16.0	–	–
Albumen	–	1.0	–	–	–	–	–	–	–
Lecithin	–	–	0.3	0.4	–	–	–	0.5	0.3
Cocoa liquor	–	–	4.7	–	–	–	8.0	11.0	–
Cocoa powder	–	–	–	5.6	–	–	–	–	–
Hazelnut	–	–	36.0	–	–	15.0	–	–	–
Flour	–	–	–	–	94.0	–	–	–	–
Salt	–	–	–	–	0.3	–	–	–	–
Soda	–	–	–	–	0.3	–	–	–	–
Dried Egg	–	–	–	–	3.4	–	–	–	–
Water	10.0	5.0	–	–	–	–	–	–	–
Total fat %	10.0	7.5	36.0	35.0	3.4	9.6	20.0	33.4	32.8

Figures give typical % compositions. FCMP, full-cream milk powder; SMP, skimmed-milk powder; SCMS, sweetened condensed milk solids.

14.6 Other factors

The complexity of confectionery products is highlighted in Table 14.13, each ingredient probably requiring a separate chapter to do this topic full justice. The main factors influencing flavour stability of confectionery products have however been covered. There is always the potentially 'hidden' factor, such as odour arising from printing inks which can on occasion give off-flavours similar to those arising from oxidative rancidity. The great measure of success achieved by the confectionery industry can be related to the use of components which offer mutual protection against autoxidation and the control of water activity.

Acknowledgement

Bob Howker's assistance in identifying the relevant literature and the support of Loders Croklaan are gratefully acknowledged.

References

1. Ames, J. M. (1988). *Chem. Ind. (London)*, 558.
2. Schmidt, R. H. (1990). In *Bitterness in Foods and Beverages*, (ed. R. L. Roussef), Elsevier Science, The Netherlands, p. 193.
3. Frankel, E. N. (1991). *J. Sci. Food Agric.*, **54**, 495.
4. Labuza, T. P. (1971). *CRC Crit. Rev. Food Technol.*, **2**, 355.
5. Meshedani, T., Pokorny, J., Davidek, J. and Panek, J. (1990). *Die Nahrung*, **34**, 915.
6. Padley, F. B. and Willems, M. G. A. (1985). *J. Am. Oil Chem. Soc.*, **62**, 456.
7. Patterson, H. B. W. (1989). *Handling and Storage of Oilseeds, Fats and Meal*, Elsevier Applied Science, London, New York.
8. Carpenter, A. P. (1979). *J. Am. Oil Chem. Soc.*, **56**, 668.
9. Muhler-Mulot, W. (1976). *J. Am. Oil Chem. Soc.*, **53**, 732.
10. Rossell, J. B. (1992). *Lipid Technol.*, **4**, Part 2, 39.
11. Gould, G. W. (1989). *Mechanism of Action of Food Preservation Procedures*, (ed. G. W. Gould), Elsevier Applied Science, London and New York, p. 97.
12. Swoboda, P. A. T. and Peers, K. E. (1977). *J. Sci. Food Agric.*, **28**, 1010.
13. Swoboda, P. A. T. and Peers, K. E. (1977). *J. Sci. Food Agric.*, **28**, 1019.
14. Swoboda, P. A. T. and Peers, K. E. (1978). *J. Sci. Food Agric.*, **29**, 803.
15. Al Shabibi, M. M. A., Langner, E. H., Tobias, J. and Tuckey, S. L. (1964). *J. Dairy Sci.*, **47**, 295.
16. Kinsella, J. E. (1969). *Chem. Ind. (London)*, 36.
17. Reimerdes, E. H. and Mehrens, H. A. (1988). *Industrial Chocolate Manufacture and Use*, (ed. S. T. Beckett), Blackie Academic & Professional, Glasgow and London, p. 47.
18. Jeon, I. J. (1993). In *Food Taints and off-flavours*, (ed. M. J. Saxby), Blackie Academic & Professional, London, p. 122.
19. Fleming, M. G. (1991). *Farm Food*, **1**, 26.
20. Shipe, W. F., Bassette, R., et al. (1978). *J. Dairy Sci.*, **61**, 855.
21. Bassette, R., Fung., D. Y. C. and Mantha, V. R. (1986). *CRC Crit. Rev. Food Sci. Nutr.*, **24**, 1.
22. Minifie, B. W. and Butt, K. C. (1968). *Manufact. Confect.*, **48**, 39.
23. Hall, G. and Lingnert, H. (1984). *J. Food Qual.*, **7**, 131.

24. Stokoe, W. N. (1928). *Biochem. J.*, **22**, 82.
25. Kinderlerer, J. L. and Kellard, B. (1984). *Phytochemistry*, **23**, 2487.
26. Kellard, B., Busfield, D. M. and Kinderlerer, J. L. (1985). *J. Sci. Food Agric.* **36**, 415.
27. Gunstone, F. D., Harwood, J. H. and Padley, F. B. (1986). *Lipid Handbook*, 2nd edn, Chapman & Hall, London, New York, p. 97.
28. Woodroof, J. G. (1979). *Tree Nuts*, AVI, Westport, Connecticut, USA.
29. Woodroof, J. G. (1983). *Peanuts*, AVI, Westport, Connecticut, USA.
30. Fourie, P. C. and Basson, D. S. (1989). *Lebensm.-Wiss. u. Technol.*, **22**, 251.
31. Holland, B., Welch, A. A., Unwin, I. D., Buss, D. H., Paul, A. A. and Southgate, D. A. T. (1991) *McCance and Widdowson, The Compositions of Foods*, Royal Society of Chemistry and MAFF (UK).
32. Keme, T., Messerli, M., Sheijbal, J. and Vitali, F. (1983). *Rev. Chocolate Confect. Baker*, **8**, (2), 15.
33. Kinderlerer, J. L. and Phillips-Jones, M. K. (1991). In Modern Methods in Food Mycology, (eds R. A. Samson, A. Hocking, J. I. Pitt and D. King) Elsevier, Amsterdam, New York, p. 123.
34. Kinderlerer, J. L. and Johnson, S. (1992). *J. Sci. Food Agric.*, **58**, 89.
35. Young, C. C. (1989). In Rancidity in Foods, (eds J. C. Allen and R. J. Hamilton), Elsevier Applied Science, London, New York, p. 179.
36. Labuza, T. P. (1970). *Food Technol.*, **24**, 543.
37. Rockland, L. B. and Nishi, S. K. (1980). *Food Technol.*, **34** (4), 42.
38. Eichner, K. (1986). In *Food Packaging and Preservation*, (ed. M. Mathlouthi), Elsevier Applied Science, London, p. 67.
39. Labuza, T. P. (1980). *Food Technol.*, **34** (4), 36.
40. Minifie, B. W. (1989). *Chocolate Cocoa and Confectionery*, 3rd edn, Van Nostrand Reinhold, p. 301.

15 Rancidity in fish
P. HARRIS and J. TALL

15.1 Introduction

Earlier chapters described how the number of double bonds in a fatty acid affected its susceptibility to oxidation; the higher the number of double bonds, the more prone to rancidity. It is well known that both fresh and saltwater fish contain significant levels of polyunsaturated fatty acids, particularly eicosapentaenoic (ω^3 20:5) and docosahexaenoic (ω^3 22:6) acids. Consequently, such fish should readily go rancid on storage.

The development of rancidity at ambient or chill storage temperatures is not a major problem with fish because the normal level of microbial flora associated with fresh fish is such that bacterial spoilage will render it inedible before rancidity has proceeded to any great extent. However, rancidity development is very important in the storage stability of dried or frozen fish.

All fish do not contain the same level of fat; the total level of fat (or lipid) in a fish will vary depending on the species of fish. Fish can be simply divided into low and high fat species. Low fat species usually have less than 2% total fat and high fat species more than 5%. In a typical fish, about 1% of its total body weight will consist of various phospholipids which are distributed throughout the body and perform essential functions such as regulating the properties of membranes. Any lipid in addition to this level will be made up of triacylglycerides which are primarily deposited in the liver and under the skin and act as a food reserve.

Total fat level is not only dependent on the species of fish, but will vary also according to the season or time of the year. Seasonal variation in fat level is due the availability of food and, in mature fish, the state of the breeding cycle; during spawning, most of the fat (and some of the muscle tissue) is converted into reproductive tissue. In some species the seasonal variation in fat content can be very marked; for example in Atlantic herring (*Clupea harengus*), it can vary from 1 to 25%.

Thus it can be seen that fish contain varying levels of total fat (depending on species and the season) and that this fat can contain highly unsaturated fatty acids, all of which can make fish very susceptible to rancidity. The purpose of this chapter is to summarise what is known about the mechanism(s) of rancidity development in whole fish (not fish oil), what techniques have been used to measure it, and how it can be controlled.

15.2 Mechanisms

As noted above, fish may be divided into two categories depending upon the location and quantity of lipid stored in the body. 'Lean fish' such as cod (*Gadus morhua*) and haddock (*Gadus aeglefinus*), predominantly store lipid in the liver. The muscle tissues of such fish contain 0.5–1.5% lipid in the form of functional lipid such as lipoprotein or phospholipid of the cellular membranes. 'Fatty fish' however also store lipid under the skin and in the muscle tissues; this can be as much as 15% of the muscle, in the form of triacylglycerides.[1]

Fish tissue susceptibility to rancidity does not only depend on the amount of lipid present, but the lipid composition and its location in the fish tissue matrix.[2] Fish contain high levels of highly unsaturated fatty acids and it is for this reason that they are susceptible to oxidative rancidity. Other lipid-type materials, such as tocopherols and carotenoid pigments are also associated with autoxidative reactions in fish tissue. These may be involved in controlling oxidative rancidity, or may be oxidized themselves or co-oxidized with lipids.[2]

Much of the information concerning lipid oxidation processes in fish has been derived from the study of oil or simple model systems. Fish, however, comprise a complex heterogenous material where lipids are found to exist in totally different environments.

Although the process of lipid oxidation is highly favourable thermodynamically, the direct reaction between oxygen and even highly unsaturated lipids is kinetically hindered.[2] Hence, an activating reaction is necessary to initiate free radical chain reactions. Investigation of the initiation process has proved to be quite difficult as only a few of the initiating species are usually generated before propagation commences. This is particularly the case for fish as the intermediate hydroperoxides are extremely unstable, due to their highly polyunsaturated nature, and break down almost instantaneously. It has been proposed that lipid oxidation in fish may be initiated and/or promoted by a number of mechnisms including the production of singlet oxygen; enzymatic and non-enzymatic generation of partially reduced or free radical oxygen species (i.e. hydrogen peroxide, hydroxy radical); active oxygen iron complexes and thermal- or iron-mediated homolytic cleavage of hydroperoxides.[3]

15.3 Non-enzymic initiation

The exact mechanisms whereby lipid oxidation is non-enzymically initiated in fish have yet to be established. However, a number of possible mechanisms can be deduced by reference to other biological systems.

15.3.1 Formation of active oxygen species

Hydroperoxides can be formed following exposure to light of a photosensitizer in the presence of oxygen. During this mechanism, singlet oxygen is formed which reacts directly with carbon–carbon double bonds of unsaturated fatty acids to form lipid peroxides. Types of photosensitizers present in fish include flavins, retinal and haem pigments.[3]

The production of active oxygen species such as superoxide anion and hydroxy radicals are also thought to be of importance in the initiation and promotion of rancidity in most biological tissues. In the cytosol of a cell, $O_2^{\cdot-}$ (superoxide anion radical) may be produced via the action of several oxidases. For example, cytochrome oxidase catalyses the transfer of electrons from cytochrome c to oxygen in the electron transport chain. Usually four electrons are transferred resulting in the production of two water molecules.[4] However, partial reduction of the oxygen generates a superoxide anion (reaction 1):

$$O_2 + e^- \rightarrow O_2^{\cdot-} \tag{1}$$

Although the cytochrome oxidase mechanism theoretically should not release the superoxide anion, a small amount is unavoidably formed. Autoxidation of oxy-myoglobin and oxy-haemoglobin (both in the Fe^{2+} oxidation state) may also result in the formation of superoxide anion and met-myoglobin and met-haemoglobin, respectively (both in the Fe^{3+} oxidation state).

Free iron can behave as a free radical and can take part in electron transfer reactions with molecular oxygen which may also lead to the generation of a superoxide anion (reaction 2):

$$Fe^{2+} + O_2 \rightarrow (Fe^{2+}\!-\!O_2 \leftrightarrow Fe^{3+}\!-\!O_2^-) \rightarrow Fe^{3+} + O_2^{\cdot-} \tag{2}$$

The superoxide anion is insufficiently reactive to abstract hydrogen from lipids. It would also be unlikely to enter the hydrophobic interior of the membrane because of its charged nature. However, the superoxide anion may be protonated to give a hydroperoxyl radical ($HO_2^{\cdot}$) (capable of initiating autoxidation) which can react with another hydroperoxyl radical to form hydrogen peroxide (reaction 3):

$$O_2 \xrightarrow{e^-} O_2^{\cdot-} \xrightarrow{H} HO_2^{\cdot} \xrightarrow{HO_2\ O_2} H_2O_2 \qquad (3)$$
$$\text{superoxide} \quad \text{hydroperoxyl} \quad \text{hydrogen}$$
$$\text{anion} \quad \text{radical} \quad \text{peroxide}$$

Alternatively, the superoxide anion may react with H_2O_2 to form an OH^- radical; the reaction requires a metal catalyst in its reduced state[5] (reaction 4):

$$H_2O_2 + O_2^{\cdot-} \rightarrow \text{metal catalyst} \rightarrow O_2 + {}^{\cdot}OH + OH^- \qquad (4)$$
$$(\text{e.g. } Fe^{2+})$$

Simple iron complexes (Fe^{2+}. ADP and salts) can react with H_2O_2 and/or $O_2^{\cdot-}$ to form hydroxy radicals (reaction 4). Iron is able to react with hydrogen peroxide via the Fenton–Haber–Wiess reaction[3,6] (reaction 5):

$$Fe^{2+} + H_2O_2 \rightarrow Fe^{3+} + OH^- + OH^{\cdot} \qquad (5)$$

Hydrogen peroxide may also be produced via peroxisomal oxidases and mitochondrial electron transport (with the aid of a metal catalyst). Both the hydroxy radical and hydroperoxide are membrane permeable and thus may come into contact with the fatty acid regions of membranes.[7]

Hydroxy radicals may also be formed in tissues by the homolytic fission of hydroperoxide groups on lipid molecules catalysed by heat, radiation and the Fenton–Haber–Wiess reaction. The hydroxy radical initiates autoxidation by abstracting hydrogen from another lipid radical thus propagating the chain reaction.

15.3.2 Activated haem proteins

It is generally accepted that intact iron proteins do not react with H_2O_2 or O_2^- to form hydroxy radicals unless the iron is released from the protein. Thus, initiation by myoglobin and haemoglobin may involve the release of iron from these proteins.[7] Intact haem proteins are not true initiators but may catalyse the propagation step by the homolytic scission of hydroperoxide group of the lipids. However, H_2O_2 can activate haem proteins so that they become true initiators. Here, H_2O_2 is thought to activate the prosthetic ferric haem group to a higher redox state ($P^+\mathrm{-Fe}^{4+}\mathrm{=O}$), which can then attack a lipid as a true initiator.[8]

Met-myoglobin in the presence of hydrogen peroxide increases lipid oxidation although neither hydrogen peroxide nor met-myoglobin alone has much effect. Although the presence of hydrogen peroxide induces the production of non-haem iron, it is thought that this is not the main cause of increased oxidation, rather met-myoglobin is activated by hydrogen peroxide as described above.[9] Hydrogen peroxide generated by a glucose oxidase/glucose system can also activate met-myoglobin to initiate lipid peroxidation of muscle microsomes.[10]

15.3.3 Non-enzymic initiation in fish

Evidence for the existence of at least two non-enzymic catalysts of lipid oxidation in fish was found when the press juice from ordinary muscle of mackerel (*Scomber scombrus*) was examined. The below 5 kDa fraction of the juice contained iron complexes whereas the above 5 kDa fraction probably contained H_2O_2 activated haem proteins.[11] Iron in the below 5 kDa fraction must be maintained in its reduced form which may be achieved *in vivo* by reductants such as ascorbate and/or cysteine.[11]

Non-haem iron increased during frozen storage of cod and mackerel due to haem breakdown; storage at $-14\,°C$ appeared more deleterious to the haem molecule than storage at lower temperatures (-20 and $-40\,°C$).[12] Vacuum packed samples had less haem breakdown than air packed samples and ascorbic acid added as an antioxidant appeared to protect the haem molecule.[13] Similar properties were observed for the soluble fraction of mackerel muscle. During storage on ice the iron content of the below 10 kDa fraction increased. Storage of frozen-thawed muscle led to an increase in both iron and copper in the below 10 kDa fraction.[14]

15.4 Enzyme initiation

There are two forms of enzyme induced oxidation. The true type involves the enzymes lipoxygenase or cyclo-oxygenase.[3,6] These enzymes directly introduce an oxygen atom on to the carbon chain of the lipid molecule. In the other type of enzyme induction (microsomal lipid oxidation) the enzymes are involved in reducing iron complexes which are then capable of stimulating lipid oxidation[3,6,8] or react with hydrogen peroxide to form the hydroxy radicals as described above. It should not be forgotten that enzymes may also be involved in the formation of other active oxygen species such as the superoxide anion or in the production of hydrogen peroxide.

15.5 Microsomal enzyme lipid oxidation

The microsomal enzyme system requires the presence of the co-factor NADH/NADPH as well as iron. The reactions are usually enhanced in the presence of ADP or ATP.[3] In general most microsomal fractions studied prefer NADPH as a co-factor,[15,16] however, fish microsomes function better with NADH compared with NADPH.[17,18] The fish system has a relatively high activity at low temperatures compared to avian and mammalian systems which may be a function of the PUFA (poly unsaturated fatty acids) present in the fish microsomes or the influence of different antioxidants.[19,20] The Michaelis constant (Km) is a measure of affinity of an enzyme for a substrate and equals the substrate concentration at which the reaction rate is half the maximum rate. The Km value for NADH is much lower in the presence of ADP, however, the concentration of NAD^+/NADH is probably higher in the muscle than the Km value expressed without ADP.[21] The ADP's function is probably to complex with Fe^{3+} to keep it soluble so that it may be reduced by the NADH enzyme system. The addition of inorganic phosphate has an inhibitory effect as phosphate interacts with the Fe^{3+} forming an insoluble complex. The NADH-dependent enzyme system was also studied in the microsomal fraction of light and dark muscle fractions of herring. The dark muscle was found to have more activity than the white flesh even though the activation energies were similar. Both systems had a pH optimum at pH 6.7 and were inhibited at high concentrations of ADP and NADH; the ferrous ion gave greater stimulation of the enzyme reaction than the ferric form.[22]

The enzyme cytochrome P-450 has also been implicated in the induction of lipid oxidation. This enzyme is a membrane-bound haem protein that metabolizes hydrophobic foreign substances called xenobiotics as well as endogenous hydrophobic substances such as fatty acids. It is postulated that two electrons are transferred from NADPH via NADPH–cytochrome-c-reductase to cytochrome P-450. The P-450 reductase enzyme, in addition to the reducing cytochrome P-450, can donate electrons to some Fe^{III}-complex and so generate Fe^{II}, which in turn stimulates oxidation. It is considered that many factors affect the cytochrome P-450 system in fish; i.e. sex, maturation, age, the lipid composition of the membranes, accessibility to co-factors, xenobiotics, nutrition, temperature and seasonal variations.[6]

It has also been proposed that NADH dehydrogenase of the respiratory chain plays a role in lipid oxidation in the mitochondria. NADH dehydrogenase may donate electrons to ADP-chelated Fe^{III} resulting in free radical generation.[3]

15.6 True enzyme initiation

True enzyme induced lipid oxidation involves the addition of oxygen directly on to the lipid molecule. One such haem-containing enzyme, cyclo-oxygenase, generates prostanoids via the stereospecific introduction of oxygen into arachidonic acid[6] ($20:4n - 6$).

More is known about the other true enzyme, lipoxygenase (Figure 15.1), which is thought to occur in the cytosol or in the microsomal fractions and has both cytosol and membrane-bound activity. The enzyme catalyses the addition of oxygen to free fatty acids containing a 1,4-*cis*-pentadiene group to form a hydroxyl. It is not considered to be reactive with triacylglycerides or phospholipids, although recent work carried out with sardine skin lipoxygenase (*Sardinops melanosticus*) has reported higher activity with the triacylglyceride, trilinolein, than the free linoleic fatty acid.[23]

Hydroperoxides formed are reduced either enzymically (glutathione peroxidase) or non-enzymically to hydroxyl analogues.[24] Alternatively they may decompose to give free radicals that can initiate chain reactions to progress autoxidation.

Most of the work to date has investigated the presence of lipoxygenase in the gill and skin tissues of various fish.[24–28] The gill system has been used extensively to investigate lipoxygenase action in fish because it is thought to be relatively free from other oxidizing reactions.[27] There has been little work reporting the presence of lipoxygenase in fish flesh, in fact Hsieh *et al.* found no detectable lipoxygenase activity in the muscle tissues of a number of freshwater fish.[28] However, Wang *et al.*[29] found

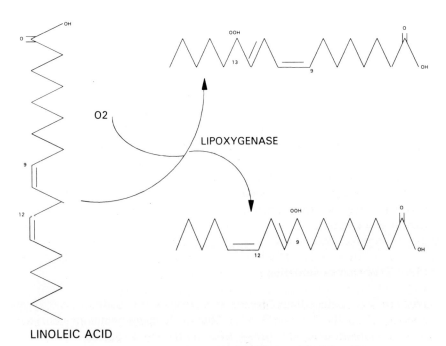

Figure 15.1 Lipoxygenase-catalysed oxidation of linoleic acid.

lipoxygenase activity was higher in the light muscle than in the skin and dark muscle of lake herring. Josephson *et al.* used whole fish to provide evidence for the involvement of lipoxygenase in producing volatiles associated with fresh fish aroma of both fresh and saltwater fish.[30] There is also work to suggest that the volatiles derived enzymically are different from those derived via autoxidative routes of freshwater fish.[31]

The evidence suggesting that lipoxygenase plays an important role in both the onset of rancidity as well as contributing towards fresh fish aroma has grown in recent years. To date, two different lipoxygenases have been discovered in fish tissue[31] but it is not yet known what tissues in addition to skin and gills express these enzymes. It is quite possible that there are distinct tissue and cell-type differences in the enzyme's distribution and that different lipoxygenases may produce different volatiles from the same fatty acid precursors.

In order to predict the variation in volatile production via lipoxygenase it is important to know its substrate specificity. The substrate specificity of lipoxygenase from partially purified sardine skin was reported to be for linoleic and α-linolenic fatty acids,[26] whereas that from rainbow trout skin (*Salmo gairdnerii*) appeared to prefer $n-3$ PUFAs to the $n-6$ arachidonic acid; linoleic acid was not tested.[27] We have recently shown that a lipoxygenase enzyme from mackerel flesh oxidized linoleic and docosahexaenoic acids more efficiently than eicosapentaenoic or linolenic acids.[32] Trout gill showed high substrate preferences for eicosapentaenoic acid, docosahexaenoic acid and arachidonic acid with only a little reactivity for linoleic acid.[28] Soyabean lipoxygenase has been used as a model system to study the hydroperoxides formed from $n-3$ and $n-6$ fatty acids and to relate the type of off-flavour produced in fish by lipoxygenase mediated lipid oxidation.[27] It was shown that, of the fatty acids tested, the $n-3$ fatty acids were the preferred substrates and that there was only one major hydroxy analogue formed for each of the fatty acids. GC–MS gave further evidence to show that the hydroxy group is located at the C-15 position of arachidonic acid and eicosapentaenoic acid, at C-13 of linolenic acid and C-17 of docosahexaenoic acid. This showed that soyabean lipoxygenase II specifically catalyses oxygen attack on the $n-6$ carbon of fatty acids.[27] Lipoxygenase activity in trout skin tissue was crudely extracted and incubated with radioactive ^{14}C arachidonic acid. 12-Hydroxyarachidonic acid was identified as the main product indicating that trout skin lipoxygenase is $n-9$ specific.[27] German and Kinsella[25] also used a crude extract from fish skin as a source of lipoxygenase for incubation with docosahexaenoic and arachidonic acid. The positional specificity of the enzyme determined by GC–MS confirmed again the presence of 12-lipoxygenase, i.e. it was $n-9$ specific. There was also some evidence to suggest that another hydroxy fatty acid derivative was formed implicating the presence of another lipoxygenase.

Mohri et al.[26] demostrated the presence of a lipid peroxidation enzyme in sardine skin. The pro-oxidant was found to have typical properties of lipoxygenase, i.e. it oxidized fatty acid stereospecifically, but was also shown to be a haem-containing protein when isolated by gel electrophoresis after partial purification by gel chromatography. Most lipoxygenases contain iron at their active site but do not contain a haem structure.

The presence of an $n-9$ lipoxygenase in gill tissue of both marine and freshwater fish has been confirmed.[33] This enzyme appears to have a common stereospecificity to that of mammalian lung tissue lipoxygenase in that it will react with fatty acids containing double bonds at $n-9$ and $n-12$ positions to produce 12(S) hydroperoxy fatty acid isomers. However, moderate activity was observed with linolenic acid which again indicates the presence of another lipoxygenase. During purification of the $n-9$ enzyme an additional lipoxygenase with $n-6$ activity was discovered. It was shown that the $n-9$ lipoxygenase requires double bonds at $n-6$ and $n-12$, whilst the $n-6$ lipoxygenase requires double bonds at $n-6$ and $n-9$ positions.[34] The $n-6$ lipoxygenase was difficult to detect in the crude gill extract and it was postulated that the extract exhibited an inhibitor for the $n-6$ lipoxygenase activity as the Kms of both lipoxygenases were identical.[34]

Oxygenation in trout gill was inhibited by the presence of esculetin (a known inhibitor of mammalian lipoxygenase), stannous chloride and by boiling the extract. The presence of ethyl glycol bis(aminoethyl ether) $N,N,-N',N'$-tetra-acetic acid (EGTA), a potent metal chelating agent, also reduced activity suggesting that a metal co-factor is a requirement for the enzyme.[25] Josephson et al. also demonstrated that the production of lipoxygenase-derived volatiles was inhibited by the presence of esculetin in whole fish. They also showed that, although thermal exposure inactivated lipoxygenase, it had the effect of enhancing the rate of autoxidation reactions in fish homogenates.[31]

Instability of the enzyme has been observed on a number of occasions,[25,28,35] and has been attributed to the accumulation of excess amounts of hydroperoxides. Stability is improved by adding glutathione which removes the hydroperoxides,[28,35] however, low levels of hydroperoxides are required for lipoxygenase activity.[28] In the absence of glutathione, 50% of the hydroperoxides are converted non-enzymatically to trihydroxy derivatives whereas the $n-9$ monohydroxy derivative is produced when glutathione is present.

Boiled extract from trout gill showed no lipoxygenase activity, however, high activity was recovered at temperatures near freezing.[28] At 40 °C, 80% of activity was retained in trout gill extract. At 60 °C and above only minimal activity was recovered from the trout gill tissue and 90% of the activity was lost after the extract had been exposed to 60 °C for 20 s.[28] Wang et al.[29] also found that heat treatment inactivated lipoxygenase activity in lake herring. However, non-enzymatic oxidation appeared to be accelerated during subsequent storage after heat activation.

15.7 Measurements

15.7.1 Chemical measurements

The two most common chemical methods used to monitor rancidity in fish are the peroxide value (PV) and the thiobarbituric acid (TBA) tests. However, these methods were originally developed to measure rancidity in oil/fat systems but when they are translated to foods they often lose sensitivity and reliability (see chapters 3 and 6).

PV determination may be of less value in assessing the quality of the highly unsaturated fats in fish because the peroxides initially formed react very quickly to form the secondary oxidation products. Thus the amount of peroxides remains low relative to that which would be formed from vegetable oils or animal fats, even after extensive oxidation.

There are two basic methods for carrying out the TBA test, i.e. either directly on the food product, followed by extraction of the coloured pigment, or on a steam distillate prepared from the food and subsequently reacted with the reagent.

Although PV and TBA determinations have their limitations, they can be useful indicators of rancidity and help to distinguish between the effectiveness of different treatments. Tomás and Añón[36] applied the TBA test to both the steam distillates and to the meat exudates obtained by centrifugation of salmon. They used this test to determine whether the cellular damage to muscle fibres caused by slow freezing resulted in increased lipid oxidation on frozen storage at $-5\,°C$, a temperature at which unsaturated fatty acid oxidation should be enhanced (compared to the control at $-25\,°C$). The results indicated that the slow freezing of any part of salmon did not increase lipid oxidation.

Santos and Regenstein[37] used PV and TBA to determine the shelf life of frozen hake (*Urophycis tenuis*) and mackerel and investigated the use of vacuum packaging, glazing and erythorbic acid.

Hwang and Regenstein[38] used PV to measure rancidity development in minced menhaden (*Brevoortia patronus*, *Brevoortia tyrannus*) treated with various antioxidants and synergists. In a storage trial at $-7\,°C$ and $-20\,°C$ they were able to monitor the performance of each antioxidant tested by determining the PVs. The effect of immersing sardines (*Sardinella longiceps*) in various concentrations of salt solutions was investigated by Nambudiry.[39] He measured the development of rancidity during the storage of the fish at $-5\,°C$ for 10 days by using TBA and the PV tests as well as the development of free fatty acids to monitor lipid oxidation; the PV and TBA tests showed similar development values.

15.7.2 Physical measurements

Several spectrometric techniques have been developed to give an estimation of the extent of rancidity in fats. When fatty acids are oxidized to give their hydroperoxides, the double bonds in many of these fatty acids become conjugated. The conjugated acids absorb UV light at 232 nm as do some of the conjugated dienes that may result from decomposition. Other secondary oxidative products absorb elsewhere in the UV range, for example the diketones show an absorbance band at 268 nm. Infrared spectroscopy has also been used to measure rancidity in oils and shown to be of value in the recognition of unusual functional groups and in the study of fatty acids with *trans*-double bonds. These spectrometric techniques are more valuable when they are used in conjuction with HPLC. A number of workers such as Perkins and Pinter[40] and Neff *et al.*[41] have used reverse phase HPLC followed by GC–MS to investigate a number of non-volatile end-products of lipid oxidation.

Techniques based on fluorimetry have been used to measure lipid oxidation mainly in biological tissues. Smith *et al.*[42] used fluorescence to measure lipid oxidation in Indonesian salted-dried marine catfish (*Arius thalassinus*). It has been shown that when lipid oxidation occurs in food such as salted dried fish, further reactions are possible between the primary and secondary lipid oxidation products and other, mainly nitrogenous food constituents, e.g. protein, amino acids and some phospholipids to give tertiary products and the possibility of browning. Measurement of fluorescence derived from the interactions between carbonyl compounds and the amino group of amino acids together with measurement of levels of acetic acid soluble colour indicate the extent of the tertiary reactions and hence lipid oxidation.

Gas chromatography (GC) is becoming increasing more popular to measure oxidative rancidity in oils and foods. Methods are being developed that determine the volatile components (the secondary products) produced by lipid oxidation. These methods of volatile analysis are known as headspace analysis. GC analysis is also used for the characterisation of lipids and some workers have investigated the use of fatty acid profiling to measure rancidity. Beltràn and Moral[43] carried out GC analysis to measure the fatty acid profile of sardine during frozen storage.

Analysis of volatiles by GC has the advantage that the concentrations of the volatile components are measured because it is some of these volatile molecules which constitute the rancid odour. However, although the peak areas are proportional to the mass of each component the palate response is dependent on the chemical structure of each molecule. Only a few researchers have looked into analysis of headspace volatiles and here again much of the work has been carried out on oils.

Matiella and Hsieh[44] analysed crab meat volatiles. The crab meat was homogenized in a solution containing an internal standard, this homogenate was then purged with pure helium and volatiles were absorbed onto a Tenax trap. This trap was then flash heated and a carrier gas used to sweep the desorbed volatiles to be cryogenically concentrated at the head of the capillary GC column. GC analysis was used to compare the volatile flavour components of boiled and pasteurized blue crab and it was found that boiled crab meat contained higher levels of total volatiles than pasteurized.

Hsieh and Kinsella[24] looked at the generation of specific volatiles by lipoxygenase. They prepared a crude enzyme extract from fish gill tissue. 2-Nonanone was added to this preparation as an internal standard after which aliquots were incubated with radiolabelled fatty acids. During incubation, samples were purged with nitrogen and volatiles were collected on a Tenax trap which was later extracted with hexane. A scintillation count as well as GC–MS analysis were then carried out on this extract. It was found that volatile formation depended upon duration of incubation, protein and substrate concentration. Hsieh et al. have also used dynamic headspace methods to determine the volatile components of menhaden fish oil.[45] The oil was purged with high purity helium 50 ml min^{-1} at 65 °C for 1 h onto a polymer Tenax bed. The volatiles were then desorbed for 20 min at 185 °C and cryogenically focused with dry ice–methanol into fused silica capillary columns. The volatiles were then assessed by GC–MS. The GC was also equipped with a sniffing port so that the eluting volatiles could be assessed by smell.

15.8 Control of rancidity in fish

Before considering the control of rancidity development in fish, it is necessary first to ask the question why you want it controlled. The answer is invariably to extend the length of time that the fish remains edible. The need for the fish (or fish product) to be edible obviously limits the use of rancidity control agents or treatments to those which are 'food grade'.

The effects of rancidity development can be totally eliminated by retorting or canning the fish. The combination of high temperatures and the total elimination of oxygen during the canning process ensures that even high fat fish such as sardines or mackerel remain edible for years. This is not to say that there is no rancidity development in the fish before they are canned because almost certainly there is, but that the stage of rancidity is insufficiently advanced for significant levels of the products of rancidity to have accumulated (canning then inhibits further rancidity development). The disadvantage of this approach is that the canning process causes significant changes to the texture, flavour and appearance of the fish such that the canned product bears little resemblance (in sensory terms) to the starting fish.

The complete elimination of rancidity development in fresh or frozen fish is not yet possible. There are a number of reasons for this, the main one being that the mechanism of rancidity development in fish flesh is still not yet fully understood. Although rancidity in fish cannot be completely eliminated, there are a number of reported steps that can be taken to minimise the speed of rancidity development.

15.8.1 Storage temperature

The lower the storage temperature the slower the rate of rancidity development; storage temperatures as low as −30 to −40 °C are needed in order to achieve a significant increase in storage life for most fish species.[46] It is worth remembering that the rate of rancidity is higher at about −5 °C than it is at 0 °C. This is because the formation of ice crystals causes the reactants remaining in solution to become more concentrated and thus stimulate the reactions to go faster (even though the temperature is lower). As the temperature reduces below −5 °C, the lowered temperature will dominate the reaction rate (even though the reactants may have become even more concentrated).

15.8.2 Degree of butchery

In any fish, the rate of rancidity development will be greater in minced flesh than in fillet, which in turn is greater than in whole fish. There are two probable explanations for this, first that butchery will destroy tissue integrity and allow cell contents (including initiators of rancidity) to mingle intimately and react. Secondly, disrupted tissue will have a larger surface area thus affording greater exposure to oxygen which is a vital reactant in rancidity development.

15.8.3 Oxygen control

Since the presence of oxygen is vital for rancidity to proceed, a number of steps can be taken to limit the level of oxygen.

15.8.3.1 Water glaze.
Frozen fish or fish portions are dipped in water to give a thin coating of ice which slows down the rate of oxygen permeation. Antioxidants are sometimes added to the water used for glazing.

15.8.3.2 Vacuum packaging. Oxygen level is significantly reduced if the fish is vacuum packed in gas impermeable packaging.[37] Although vacuum packaging does not prevent rancidity development, it is probably the single most effective treatment for reducing rancidity in frozen fish.

15.8.3.3 Modified atmosphere packaging (MAP). Various combinations of carbon dioxide, nitrogen and air have been used to extend the shelf life of fresh fish.[47] Fey and Regenstein[48] reported that the shelf life of hake and salmon could be extended to almost one month by storing the fish in 1% potassium sorbate ice in a MAP consisting of 60% CO_2, 20% O_2 and 20% N_2. The extension of shelf life by 50% for a number of species of fish has been demonstrated when the fish have been packed in an atmosphere of CO_2.[49]

The main effect of MAP is to slow down the rate of microbial growth, but any reduction in oxygen concentration will also slow down rancidity development.

15.8.3.4 Antioxidants. Since lipid oxidation is a chain process, there are two mechanistically distinct classes of antioxidant which can be used. The first operates by a radical chain-breaking mechanism by inactivating alkyl peroxy and alkyl radicals which are two important radical species in the chain-propagating step. The other involves the prevention of the introduction of chain-initiating radicals into the system. Because chain-breaking and preventive antioxidants interfere at different points in the oxidation process, they mutually reinforce each other resulting in synergism.[50] The two main classes of antioxidant are the phenolics such as butylated hydroxyaniline (BHA), butylated hydroxytoluene (BHT) and butylated hydroxyquinone (TBHQ) and the tocopherols (see chapter 5 for more details).

Zama *et al.*[51] reported that the rate of oxygen consumption in fish decreased as the concentration of added α-tocopherol was increased. Treatment of raw or whole sea bream (*Pagrus major*) and mackerel with tocopherol prevented lipid oxidation. Ascorbic acid may act as an antioxidant or pro-oxidant because of its *ortho* dihydroxyl group. Ascorbic acid acted as a pro-oxidant at concentrations above 500 ppm and as an antioxidant below this concentration in the dark flesh of mullet (*Mugil cephalus*).[52] Ke *et al.*[46] found that the order of effectiveness of antioxidants for inhibiting the oxidation of skin lipids in mackerel was TBHQ > BHA > BHT.

Although it can be seen that antioxidants can have some effect on inhibiting rancidity, their effects can at times be unpredictable. Recent research has focused on the antioxidation capabilities of natural flavonoids such as Quercetin, Kaempferol and Myrecitin.[53]

15.9 Conclusions

Oxidative rancidity is probably more important in fish than in any other food due to the high level of PUFAs and the number of potential initiators/promoters present. Fish are not physically very robust, therefore, the very act of catching and handling can easily lead to extensive flesh damage which will enhance deteriorative changes. Also, because fish are cold blooded, subsequent storage on ice will not reduce the rates of deteriorative changes to the same extent that low temperature storage of mammalian tissue would have.

Compared to oils and fats, the mechanisms of rancidity development in fish have not been extensively studied and much work has yet to be done before these mechanisms are fully understood.

References

1. Castell, C. H. (1971). Metal-catalyzed lipid oxidation and changes of proteins in fish. *J. Am. Oil Chem. Soc.*, **48**, 645
2. Labuza, T. P. (1971). Kinetics of lipid oxidation in foods. *CRC Crit. Rev. Food Technol.*, October, 356.
3. Kubow, S. (1992). Routes of formation and toxic consequences of lipid oxidation products in foods. *Free Radical Biol. Med.* **12**, 63–81.
4. Stryer, L. (1988). Oxidative phosphorylation. In *Biochemistry*, 3rd edn, W. H. Freeman, NY, Chap. 17, p. 397.
5. Hsieh, R. J. and Kinsella, J. E. (1989). Oxidation of polyunsaturated fatty acids: mechanisms, products, and inhibition with emphasis on fish. *Adv. Food Nutr. Res.*, **33**, 233.
6. Ingemansson, T. (1990). Lipid Metabolism and post mortem changes in fish. SIK-Raport 571, Swedish Institute for Food Research, Göteburg.
7. Halliwell, B. and Gutteridge, J. M. C. (eds) (1989) Lipid peroxidation a chain radical reaction. In *Free Radicals in Biology and Medicine*, Chap. 4, Clarendon Press, Oxford.
8. Kanner, J., German, J. B. and Kinsella, J. E. (1987). Initiation of lipid peroxidation in biological systems. *CRC Crit. Rev. Food Sci. Nutr.*, **25**(4), 317–364.
9. Rhee, K. S. (1988). Enzymic and non-enzymic catalysis of lipid oxidation in muscle foods. *Food Technol.*, 127.
10. Harel, S. and Kanner, J. (1985). Muscle membrane lipid peroxidation by H_2O_2-activated metmyoglobin. *J. Agric. Food Chem.*, **33**, 1188–1192.
11. Decker, E. A. and Hultin, H. O. (1990). Nonenzymic catalyst of lipid oxidation in mackerel ordinary muscle. *J. Food Sci.*, **55**, 951–953.
12. Gomez-Basauri, J. V. and Regenstein, J. M. (1992). Processing and frozen storage effects on the iron content of cod and mackerel. *J. Food Sci.*, **57**(6), 1332–1336.
13. Gomez-Basauri, J. V. and Regenstein, J. M. (1992). Vacuum packaging, ascorbic acid and frozen storage effect on heme and nonheme iron content of mackerel. *J. Food Sci.*, **57**(6), 1337–1339.
14. Decker, E. and Hultin, H. O. (1990). Factors influencing catalysis of lipid oxidation by the soluble fraction of mackerel muscle. *J. Food Sci.*, **55**(4), 947–953.
15. Hochstien, P. and Ernester, L. (1963). ADP-activated lipid peroxidation coupled to the NADPH oxidase system of microsomes. *Biochem. Biophys. Res. Comm.*, **12**, 388.
16. Lin, T. S. and Hultin, H. O. (1976). Enzymatic lipid peroxidation in microsomes of chicken skeletal muscle. *J. Food Sci.*, **41**, 1488.
17. McDonald, R. E., Kelleher, S. D. and Hultin, H. O. (1979). Membrane lipid oxidation in a microsomal fraction of red hake muscle. *J. Food Biochem.*, **3**, 125.

18. Hultin, H. O., McDonald, R. E. and Kelleher, S. D. (1982). Lipid oxidation in fish muscle microsomes. In *Chemistry and Biochemistry of Marine Food Products*, (eds R. E. Martin, G. J. Flick, C. E. Hebard and D. R. Ward), Avi Publishing, Westport, CT, USA, Chap. 1, p. 1, G4.
19. Apgar, M. E. and Hultin, H. O. (1982). Lipid peroxidation in fish muscle microsomes in the frozen state. *Cryobiology*, **19**, 154.
20. McDonald, R. and Hultin, H. O. (1987). Some characteristics of the enzyme lipid peroxidation system in the microsomal fraction of flounder skeletal muscle. *J. Food Sci.*, **52**(1), 15–27.
21. Murat, M. and Sakaguchi, M. (1986). Storage of yellowtail (*Seriola quinqueradiata*) white and dark muscle in ice: changes in content of adenine nucleotides and related compounds. *J. Food Sci.*, **51**, 321.
22. Slabyj, B. and Hultin, H. O. (1982). Lipid peroxidation by microsomal fraction isolated from light and dark muscles of herring (*Clupea harengus*). *J. Food Sci.*, **47**, 1395.
23. Mohri, S., Cho, S., Endo, Y. and Fujimoto K. (1992). Linoleate 13(S)-lipoxygenase in sardine skin. *J. Agric. Food Chem.*, **40**, 573–567.
24. Hsieh, R. J. and Kinsella, J. E. (1989). Lipoxygenase generation of specific volatile flavour carbonyl compounds in fish tissues. *J. Agric. Food Chem.*, **37**(2), 279.
25. German, J. B. and Kinsella, J. E. (1985). Lipid oxidation in fish tisue. Enzymatic initiation via lipoxygenase. *J. Agric. Food Chem.*, **33**(4), 680.
26. Mohri, S., Cho, S., Endo, Y. and Fujimoto, K. (1990). Lipoxygenase activity in sardine skin. *Agric. Biol. Chem.*, **54**(7), 1889.
27. Hsieh, R. J. and Kinsella, J. E. (1986). Lipoxygenase catalyzed oxidation of $n-6$ and $n-3$ polyunsaturated fatty acids; relevance to and activity in fish. *J. Food Sci.*, **51**(4), 940.
28. Hsieh, R. J., German, J. B. and Kinsella, J. E. (1988). Lipoxygenase in fish tissue: some properties of the 12-lipoxygenase from trout gill. *J. Agric. Food Chem.*, **36**, 680–685.
29. Wang, Y. J., Miller, L. A. and Addis, P. B. (1991). Effect of heat inactivation of lipoxygenase on lipid oxidation in lake herring. *J. Amer. Oil Chem. Soc.*, **68**(10), 752–757.
30. Josephson, D. B., Lindsay, R. C. and Stuiber, D. A. (1984). Variations in the occurrences of enzymically derived volatile aroma compounds in salt- and freshwater fish. *J. Agric. Food Chem.*, **32**, 1344–1347.
31. Josephson, D. B., Lindsay, R. C. and Stuiber, D. A. (1987). Enzyme hydroperoxide initiation effects in fresh fish. *J. Food Sci.*, **52**(3), 596–599.
32. Harris, P. and Tall, J. (1994). Substrate specificity of mackerel flesh lipoxygenase. *J. Food Sci.*, in press.
33. Winkler, M., Pilhofer, G. and German, J. B. (1991). Sterochemical specificity on the $n-9$ lipoxygenase of fish gill. *J. Food Biochem.*, **15**, 437–448.
34. German, J. B. and Creveling, R. K. (1990). Identification and characterization of a 15-lipoxygenase from fish gills. *J. Agric. Food Chem.*, **38**, 2144–2147.
35. German, J. B. and Kinsella, J. E. (1986). Hydroperoxide metabolism in trout gill tissue: effect of glutathione on lipoxygenase products generated from arachidonic acid and docosahexaenoic acid. *Biochim. Biophys. Acta*, **879**, 378–387.
36. Tomás, M. C. and Añón, M. C. (1990). Study on the influence of freezing rate on lipid oxidation in fish (salmon) and chicken breast muscle. *Int. J. Food Sci. Technol.*, **25**, 718–721.
37. Santos, E. E. M. and Regenstein, J. M. (1990). Effects of vacuum packaging, glazing and erythorbic acid on the shelf-life of frozen white hake and mackerel. *J. Food Sci.*, **55**(1), 64.
38. Hwang, K. T. and Regenstein, J. M. (1988). Protection of Menhaden lipids from rancidity during frozen storage. *J. Food Sci.*, **54**(5), 1120.
39. Nambudiry, D. D. (1980). Lipid oxidation in fatty fish. The effect of salt content in the meat. *J. Food Sci. Technol.*, **17**, 176.
40. Perkins, E. G. and Pinter, S. (1988). Studies on the concentration of oxidized components of abused fats and the application of HPLC to their separation. *J. Amer. Oil Chem. Soc.*, **65**(5), 783.

41. Neff, W. E., Frankel, E. N. and Weisleder, D. (1981). High pressure liquid chromatography of autoxidized lipids: II Hydroperoxy–cyclic peroxides and other secondary products from methyl linoleate. *Lipids*, **16**(6), 439.
42. Smith, G., Hole, M. and Hanson, W. H. (1990). Assessment of lipid oxidation in Indonesian salted-dried marine catfish (*Arius thalassinus*). *J. Sci. Food Agric.*, **51**, 193–205.
43. Beltràn, A. and Moral, A. (1990). Gas chromatograpy estimation of oxidation deterioration in sardine during frozen storage. *Lebensm. Wiss. u. Technol.*, **23**(6), 499.
44. Matiella, J. E. and Hsieh, T. C. Y. (1990). Analysis of crab meat volatile components. *J. Food Sci.*, **55**(4), 962.
45. Hsieh, T. C. Y., Williams, S. S., Vejaphan, W. and Meyers, S. P. (1989). Characterization of volatile components of Menhaden fish (*Brevoortia tyrannus*) oil. *J. Amer. Oil Chem. Soc.*, **66**(1), 114.
46. Ke, P. J., Ackman, R. G., Linke, B. A. and Nash, D. M. (1977). Differential lipid oxidation in various parts of frozen mackerel. *J. Food Technol.*, **12**, 37.
47. Pedrosa-Menabrito, A. and Regenstein, J. M. (1990). Shelf-life extension of fresh fish—a review. Part II – Preservation of fish. *J. Food Qual.*, **13**, 129–146.
48. Fey, M. S. and Regenstein, J. M. (1982). Extending shelf-life of fresh wet red hake and salmon using CO_2–O_2 modified atmosphere and potassium sorbate ice at 0–1 °C. *J. Food Sci.*, **47**, 1048–1054.
49. Gray, R. J. H., Hoover, D. G. and Muir, A. M. (1983). Attenuation of microbial growth on modified atmosphere-packaged fish. *J. Food Prot.*, **46**, 610–613.
50. Scott, G. (1965). In *Atmospheric Oxidation and Antioxidants*, (ed. G. Scott) Elsevier, New York, p. 115.
51. Zama, K., Takama, K. and Mizushima, Y. (1979). Effect of metal salts and antioxidants on the oxidation of fish lipids during storage under the conditions of low and intermediate moistures. *J. Food Proc. Preserv.*, **3**, 249.
52. Deng, J. C., Matthews, R. R. and Watson, C. M. (1977). Effect of chemical and physical treatments on rancidity development of frozen mullet (*Mugil cephalus*) fillets. *J. Food Sci.*, **42**, 344.
53. King, D. L. and Klein, B. P. (1987). Effect of flavonoids and related compounds on soyabean lipoxygenase-1 activity. *J. Food Sci.*, **52**, 220.

Index

abused oils
  assay 31
accelerated tests 37–50, 64, 65
Acceptable Daily Intake *see* ADI
Acid Degree Value *see* ADV
activation energy
  autoxidation 7
active oxygen assay 45, 63, 87
acute toxicity 134
additives
  approval of 209
  definition 209
  labelling 217–218
  legislation 209
  regulations 162
  safety assessment 210
ADI 210
adipose tissue 136, 191–192
ADP 261
adulteration 203
ADV 181
  milk 186
  non-dairy creams 223
advertising 218–220
age
  peroxide levels 138
agitation
  lipolysis in milk and 181–182
alcohols
  off-flavours 17
  secondary products 16
aldehydes
  assay 31, 36–37, 118
  flavour 16
  flavour thresholds 55
  headspace analysis 121
  secondary products 15, 16
alk-2-enals
  assay 30, 31
alka-2,4-dienals
  assay 30, 31
alkanes
  assay 35–37
alkoxy radicals 14, 16
allyl radicals 9, 12
almonds
  tocopherol 249
Alpura sterilisation 225
Amadori compounds
  confectionery 231, 238

animal fats 3, 4
  polyunsaturated fatty acids in 131
animal lipoxygenase 262–265
anisidine value 29–30, 56, 118, 234
'Anoxomer' 75
antimicrobials
  BHA 90
  BHT 90
  legislation 210–211, 214–215
antioxidants 18–21, 59, 83–103, 133, 235–236
  addition to oils 99
  and microbial degradation 90
  'Anoxomer' 75
  applications of 100
  BHA 227
  BHT 188, 227
  carbonyl formation 120
  carcinogenicity 86
  casein 187
  cereals 122, 154–155
  cysteine 200
  definition 84
  fish 269
  induction period 11
  labelling 218
  lard 85, 86–88
  legislation 210–211, 216
  Maillard reaction products 227
  meat 199–200
  mechanism 87
  misconceptions 90
  myocarditis 136
  natural 152, 188
  packaging 75
  permitted concentrations 101
  polymeric 102
  potato crisps 95, 97
  powdered fats 228
  properties 90
  reason for use 85
  requirements of 85
  safety 86
  salami 124
  storage of oil 73
  synergism 89, 269
  table of properties 92, 93
  tocopherol 91, 92
  use 99
AOCS methods 27, 28, 33–35, 45, 49, 56

apparatus
  cleaning procedure 49–50
  OSI 49–50
Arrhenius plots 43, 237
arthritis 134
ascorbic acid 20, 89, 91, 134
  meat 198, 200
  milk 187
ascorbyl palmitate 91, 92, 200
assay 63
  active oxygen 45
  aldehydes 36–37, 118
  alkanes 35–37, 57
  anisidine value 29–30
  bioluminescence 35
  carbonyls 118, 124
  chemiluminescence 35, 61, 113, 116–117
  chromatographic 35–37
  cocoa butter stability 242
  comparison 47
  conductivity 59
  conjugated dienes 107–108
  COP 61, 151, 108
  correlations between methods 63
  dielectric constant 35
  differential scanning calorimetry 63
  dinitrophenylhydrazine 118, 119, 124
  esterases 24
  evaluation 54–66
  fish 265–267
  fish, UV 266
  flour 157–158
  fluorescence 35, 117, 266
  fluorescent products 124–126
  free fatty acids in dairy products 180
  GC 105, 114, 119–124, 266
  GC–MS 123, 124, 176, 266
  geometric isomers 111
  headspace 36, 58, 105–106, 121–123
  HPLC 108–114, 119, 266
  hydroxy acids 111–113
  in vivo 134
  indoxyl acetate 180, 183, 184
  induction period 41
  infra red 34, 61
  ketonic rancidity 26
  Kreis test 33
  lipase in milk powder 225
  lipase in non-dairy creams 224–225
  lipases 24
  lipolysis in dairy products 180
  lipoxygenase 34, 107, 158–159
  octanoate 33
  of bleaching 106
  OSI 49–50
  oxidation resistance 37–50
  oxidative rancidity 26–37
  oxidograph 43

  oxygen 105
  oxygen absorption 39–43
  pentane in breath 134
  peroxide value 27–29
  peroxide value in dairy products 185
  peroxides 109, 113, 114
  physical methods 34–35
  polarography 35
  purge and trap 123
  radicals 106
  Rancimat 46–49, 176
  Rancimat apparatus 47–48
  refractive index 34
  Schiff bases 124
  SDE 124
  secondary oxidation product 117–124
  spectrophotometric 106–108
  static headspace 121–123
  summary of methods 105
  Swift test 45
  TBA test 31–33
  thermogravimetric 61
  totox value 31
  triglyceride hydroperoxides 109–114
  unsaturation 60
  urine 134
  UV 63
  wheat lipase 145, 156–157
  wheat oxidation 150–151
atherosclerosis 138
audit
  HACCP 176
autoxidation
  activation energy 7
  assay 26–37
  assay methods 105
  cereals 152–153
  confectionery 234–236
  dairy products 187–188
  mechanisms 8
  nuts 252
  oleate 9
  phases 11
  protection by nitrogen 79, 80
  rates 8, 68, 234
  resistance 69
  secondary products 14–17
autoxidation
  water activity and 238
AV *see* anisidine value

babymilk
  copper in 188
  oxidation 188
bacon 196, 198
bacteria
  psychrotrophs 183
bacterial enzymes
  inactivation 183

barley
  lipoxygenase 152
BDI 181
beefburgers 200
beer 152
benzoate
  legislation 211–213
'best before' dates
  biscuits 166
BHA 19, 43, 48, 75, 88, 154, 227
  antimicrobial effect 90
  fish 269
  properties 93, 94
  synergism with BHT 89, 95
BHT 19, 43, 48, 75, 188, 227
  antimicrobial effect 90
  fish 269
  myocarditis 136
  properties 93, 95
  steam volatility 95
  synergism with BHA 89
bioluminescence
  assay 35
biscuits 153
  oats 162
  packaging 163, 173
  product acceptability 167
  product development 169
bleaching
  assay of 106
  dissolved oxygen and 77
  flour 149–150
bleaching earths 162
blending
  oils and fats 165
blood
  in milk 182
'bloom' 196
bran 141, 143
  flavour 150
  rice 147
bread-making
  oxygen and 149
British Standard Methods 24, 25, 27, 45
browning
  water activity and 239
BS 5750 207
bulk delivery
  oils and fats 165
bulk oils 76, 78
Bureau of Dairy Industry see BDI
butter
  'lactic' flavour 184
  lipolysis 179, 184
  oxidation 188
  photo-oxidation 188
'butter creams' 226
butter fat
  lipolysis 183

confectionery 242–245
buttermilk
  lipolysis 183–184
buttermilk in creams 226
butylated hydroxyanisole see BHA
butylated hydroxytoluene see BHT
butyric acid
  milk 180
buying policy 164

cancer 134, 137
  nitrosamines 198
canning 267
canola oil 110, 112
carageenans 188
caramelisation
  confectionery 231
carbon dioxide
  packaging 196
  supercritical 123
carbonyl formation
  antioxidants 120
carbonyls
  assay 118, 124
carcinogenicity
  antioxidants 86
  oxidised fat 137–138
'cardboard' flavour 152, 225
  milk 185
cardiomyopathy 136
carotenoids
  wheat 150, 152
  'carry through' 94, 97–99
casein 181, 187
caseinate 227
catalase 134, 187
cephalin
  structure 1
cereal lipases
  non wheat 146–147
cereal lipids
  autoxidation 152–153
cereal lipoxygenases
  non wheat 151–152
cereal storage
  inert atmosphere 154
cereals 141–159
  antioxidants 122, 154
  environment 172
  enzymes 142
  fatty acid composition 147
  functional properties 143
  heat-treatment 153
  hydrolytic rancidity 142–147
  lipoxygenase 143, 148–152
  lipoxygenase assay 158–159
  metal ions 152
  mixtures 155
  moisture 153, 154

cereals (cont.)
    non-enzymic oxidation 152–153
    oxidative rancidity 147–153
    processing 152
    protection against rancidity 141
    relationships of rancidity 142
    shelf life 122
    stabilisation 155
    storage 154
    tocols 152
chain mechanism 8–12
chain-breaking 19
cheese
    flavour thresholds 184
    lipolysis 184
chemiluminescence 113, 116, 117
    assay 35, 61
    milk powder 116
chlorine 201
chocolate 24
    peroxide formation in 115
cholesterol
    carcinogenicity and oxidised 138
    labelling 219
    oxidation of 138
chopping *see* comminution
chromatographic assays 35–37
chromogen 32
chronic toxicity 135
churning 188
churning efficiency
    cream 179
CIP 164
citric acid 89
claims
    labelling 218–220
cleaning
    legislation 206
    pipelines 72
closed implant cleaning 164
Clostridium botulinum 198
clotting
    peroxides and 138
coatings
    tank 70
cocoa butter
    confectionery 241–242
    phospholipids 242
    stability assay 242
cocoa powder
    lipolysis 25
coconut 26
    soapy flavour 161
coconut oil
    flavour when rancid 247
    ketonic rancidity 246–247
    lipolysis 246–247
cod liver oil 135, 136
Codex Alimentarius 204–205, 207 209, 218

cod-liver oil 131
coldpressed oils 130
collagen 191
comminution of meat 192, 195
commodity standards 205
compartmentation
    meat 192
composition
    confectionery 253
conching 241
condensed milk 245
conductivity assay 59
confectionery 230–254
    Amadori compounds 231, 238
    autoxidation in 234–236
    butterfat 242–245
    caramelisation 231
    cocoa butter 241–242
    composition 253
    dairy ingredients 242–245
    flavour 230
    ingredients 230–237
    lauric fats 246–247
    lipolysis 243–244
    Maillard Reaction 231, 238
    nuts 248–253
    oils and fats 232–237
    partially-hydrogenated oils 248
    polysaccharides 231
    product acceptability 167
    product development 169
    protein 231–232
    rancidity 171
    seeds 248–253
confectionery fats 241–248
conjugable oxidation products *see* COP
conjugated dienes
    assay 107–108
consumer 166–167
consumer protection
    legislation 217
contamination
    legislation 209
control of rancidity
    fish 267–269
controlled atmosphere packaging 85, 188, 200–201
    fish 269
    meat 194, 196
cooking 132
    fish 132
    malonaldehyde 132
    meat 132, 199
    rice 147
cooking oil
    toxicity, long term 136–137
co-oxidation
    lipoxygenase 148
    wheat products 149–150
COP assay 61, 108

flour 151
copper 69, 173, 234
　babymilk 188
　milk 186, 188
　milk processing 185
　nuts 249
copper ions 59
copper vessels
　confectionery 243
correlation
　assays 63
cost reduction 169
co-substrate
　lipoxygenase 148
crab 267
cream
　churning efficiency 179
　oxidation 188
creams 222–228
　buttermilk 226
　fat components of 223
　formulation 223
　functional properties 223
　lipolysis 223
　pasteurisation 222
　shelf life 225
　short-life 222–226
　solid 226–227
　sterilisation 222
critical control point 166, 168–169
　formulation 169
　hydrolysis factors 168–169
　identification 174
　manufacture 169
　oxidation factors 168–169
cured meat 196
cyclo-oxygenase 261
cysteine 200
cytochrome c 35, 258
cytochrome c reductase 261
cytochrome oxidase 258–259
cytochrome P-450 261

dairy ingredients
　confectionery 242–245
dairy products 179–188
　autoxidation assay 185
　autoxidation in 187–188
　composition 244
　de-emulsification 185
　flavour thresholds 184
　lipolysis 179–184
　microbial degradation 182–184
　sequestering agents 188
deactivators
　UV 20
deca-2,4-dienal 56
deca-2,4,7-trienal 243
de-emulsification
　dairy products 185

deep-fat frying see frying
degeneration
　fats 135
degumming
　dissolved oxygen 77
derivativization
　of peroxides 114
desserts 222–228
deviation procedure
　HACCP 175
diarrhoea 134, 135
dielectric constant 35
dienes
　assay 107–108
diet
　oxidised fats in 129–131
　polyunsaturated fatty acids in 131
　toxicity 137
dietary pattern
　legislation 210
differential scanning calorimetry
　assay 63
dinitrophenylhydrazine
　assay 118, 119, 124
dinitrophenylhydrazones 33
diphenyl-1-pyrenylphosphine 117
directive
　labelling 217
　Nutrition Labelling 219
directives 205
disinfection
　legislation 206
dissolved oxygen 76
　milk 186–187
　peroxide value 76
　transport 78–80
distribution
　HACCP 174
DNPH 118, 119, 124
dough 172
　oxygen 149
DPPD 118
DPPP 117
drums
　coating 73
　heating 74
　storage of oil 72–74
durability see shelf life 218
Durum wheat 152

EC 205, 207, 208
EC packaging regulation 208
EC labelling directive 217
EC proposals
　claims 219
EDTA 75, 89, 188
　legislation 216
EEA 205
EFTA 205
eggs 129

electrochemical detection 113
electron spin resonance see ESR
'Embanox' 98, 99
embryonic axis 148
EN 29 000 207
encapsulation
  powdered fats 228
'ene' reaction 12
environment
  cereals 172
  packaging 166
  risk assessment 172
enzymes
  catalase 187
  cereals 142
  cyclo-oxygenase 261
  cytochrome c reductase 261
  cytochrome oxidase 258
  cytochrome P-450 261
  free radicals 134
  glutathione peroxidase 262
  inactivation 183
  lactoperoxidase 186, 188
  lipases in milk 181–182
  meat 192–193
  microsomal 260–261
  milk 186, 188
  superoxide dismutase 187
  xanthine oxidase 186, 188
enzymic initiation 260–265
epidemiology
  oxidised fats 137–138
ESR 106
essential oils 45
esterases
  assay 24
esters 17
ethylenediamine tetra-acetic acid
  see EDTA
European Economic Area see EEA
European Economic Community see EC
European Free Trade Association see EFTA
Eurotium amstelodami 26
Eurotium spp. 246

FAME 105
FAO 204
fat
  meat 191
fat globule membrane
  milk 181–182
fat globules
  milk 181–182
fats
  composition 1
  confectionery 241–248
  melting points 197
  pyrolysis 129

spray-drying 227
fats in creams 223
fatty acid composition
  nut oils 250
fatty acids
  cereals 147
  n-3 57
  n-6 57
'fatty' fish 257
Fenton-Haber-Weiss reaction 259
FFA see free fatty acids 180
FIRA-Astell test 40–43, 59
fish
  antioxidants 269
  BHA 269
  BHT 269
  butchery 268
  canning 267
  cooking 132
  fat levels 256
  frozen storage 268
  inhibition of rancidity 267–269
  lipoxygenase 262–265
  measurement of rancidity 265–267
  mechanism of oxidation 257–265
  non-enzymic initiation 260
  oxidation 123
  packaging 268–269
  peroxide value 265
  photosensitisers 258
  polyunsaturated fatty acids in 131
  rancidity 256–270
  shelf life 265
  storage 260
  TBA test 265
  TBHQ 269
fishy taints in meat 199
fishy taste 242–243
flavour 160–162, 167, 223, 237–241
  aldehydes 55
  beany 55
  bitter 151
  bran 150
  'cardboard' 152, 185, 225
  cereals 142, 143
  cheesy 18
  citrus 16
  confectionery 230
  cream 55
  cucumber 55
  fatty 16
  flour 149
  fresh 16
  fruity 18
  goaty 18
  grassy 17
  green 16, 17, 55
  improvement 90
  ketones 18

'lactic', in butter 184
lipolysis 18
meat 131, 198-199
milky 16
musty 26
oats 151
oxidation 20
oxidised 167, 186
paraffinic 18
phospholipids 199
pungent 16
rancid coconut oil 247
roast meat 199
rubbery 236
soapy 142, 153, 160, 162, 223
stale 55, 186
storage 75
sweet 18
tallow 55
'warmed over' 199, 200
flavour reversion 167, 225
flavour score 55, 56
flavour thresholds
  cheese 184
  dairy products 184
flour
  bleaching 149-150
  deterioration assay 157-158
  fungal contamination 153
  particle size 155
  rancid flavour 149
  risk assessment 172
  tocopherol 153
  wholemeal 150
fluorescence
  assay 117
fluorescence assay
  fish 266
  milk powder 225
fluorescent products
  analysis 124-126
  salami 124
fluorimetry 35
food
  irradiation 133
  quality 238
Food and Agriculture Organisation 204
food handling
  legislation 205-206
food legislation *see* legislation
Food Safety Act 204, 209
foods
  uses of antioxidants 100
formulation
  critical control points 169
'framework' directives 205
free fatty acids
  assay in dairy products 180
  hazelnuts, on storage 251

milk 244
  titration 25
  units 180
free fatty acids in milk
  pumping 181
free radicals 134
  age 138
  enzymes 134
  injury 134
  irradiation 208
  low density lipoprotein and 138
  toxicity 138
freezer storage
  fish 268
freshness 165
frozen storage
  meat 196-197
frying 132-133
  polar material 133
frying oil
  toxicity 137
functional properties
  creams 223
  oxidation of cereals and 143
fungal contamination
  flour 153

gallates
  'ink' formation 96
  properties 93, 95
  storage 96
gas chromatography *see* GC
GC 35-37, 114
  assay 105, 119-124
  fish 266
  purge and trap 123
GC-MS 123, 124, 176
  fish 266
geometric peroxide isomers
  assay 111
germ 141
ghee 138
glutathione peroxidase 134, 262
gluten 153
grain 141
green shelling
  oats 162
Greenhouse Effect 27
grid melters 164
groundnut oil
  fatty acids 60
  flavour score 56
  iodine values 60
growth rate 135

HACCP 166-176
  audit 176
  definitions 166
  deviation procedure 175

HACCP (*cont.*)
   distribution 174
   finished product 172–173
   hygiene and 207
   ingredients 170–172
   level of concern 175
   management and 172, 176
   monitoring 174
   packaging 173
   plant design 172
   policy 175
   product development 169
   production schedule and 175
   risk categorisation 169–172
   sale 174
   shelf life 176
   storage 174
   verification 176
haem 194
haem proteins
   initiation 259
haemoglobin 258–259
haemolytic anaemia 136
ham 196, 200
hazard 166
hazard analysis 166
Hazard Analysis Critical Control Points *see* HACCP
hazards
   rancidity 167–168
hazelnuts
   free fatty acids, on storage 251
   induction period 252
   taste on storage 251
headspace analysis 36, 57, 58, 105, 106, 121-123
health hazards
   legislation 206
'healthy eating' 166
   rancidity 174–175
heart 135
heart disease
   ghee and 138
heating
   of drums 74
   pipelines 72
   process vessels 71
hedonic scale 55
Henry's Law 78
hepatomegaly 135
herring oil
   irradiation 133
hexanal 57
hexanoic acid
   milk 180
High Performance Liquid Chromatography *see* HPLC
'high-linoleic acid' milk 187
highly unsaturated oils 65

history 68
   food legislation 203–204
   Schaal test 177–178
homogenisation
   milk 181, 187
'horizontal' directives 205
HPLC
   assay 35, 119, 108–114
   correlation with PV 110, 112
   fish 266
HTST
   pasteurization 183
hydrocarbons
   assay 35–37
   secondary products 17
hydrogen peroxide 258–259
hydrogenated oils
   confectionery 248
   powdered fats 228
hydrogenation 161
hydrolytic rancidity
   cereals 142–147
   confectionery 171
   meat 192–193
hydroperoxide formation
   mechanisms 8–14
hydroperoxides 106–108
   breakdown 14–17
   structures 10
hydroperoxyl radical 258
hydroperoxyoleic acid
   breakdown 15
hydroxy acids
   assay 111–113
   purification before assay 111
hydroxy radicals 258–259
hydroxybenzoate
   legislation 211–213
hydroxyl radicals 187
hygiene
   HACCP and 207
   legislation 205–207
   principles 206–207
hygiene regulation
   legislation 206–207

IDF 180, 184
Ikan bilis 131
in vivo
   assay of oxidation 134
inactivators
   metal 20
indoxyl acetate 24
   assay 180, 183, 184
induced lipolysis 181–182
induction period 11, 38–43, 181–182, 236-237
   antioxidants 11

comparisons 42, 47
determination 39–43, 57–60
hazelnuts 252
lard 59, 87, 88
soybean oil 60
vegetable oils 41, 42
inert atmosphere
cereal storage 154
infra red
assay 34, 61–62
ingredient lists
labelling 217
ingredients
confectionery 230–237
labelling 217–218
oils and fats 161
risk assessment 170–172
inhibitors
lipase 154, 181
lipoxygenase 264
photo-oxidation 13
proteins 133
initiation 8–12, 18, 57, 258–265
cytochrome c 258
cytochrome oxidase 258
enzymic 260–265
fish, non-enzymic 260
haem proteins 259
hydroxy radicals 258
iron 258
met-myoglobin 259
non-enzymic 258–260
superoxide 258
injury
free radicals 134
ink
packaging 173
inks 96, 163
International Dairy Federation *see* IDF
intersystem crossing 12
intrinsic milk lipase 181
iodine value 60, 223
iodometric titration 27
iron 89, 234
cereals 153
haem 194–196
initiation 258
mackerel 260
nuts 249
irradiation
legislation 208
poultry 133
irradiation of food 133
ISO 9000 207
IUPAC methods 30
IV *see* iodine value

Karl Fischer titration 24
keto-acids 17

ketones 17, 18
flavours 18
headspace analysis 121
ketonic rancidity 22, 26, 168, 246–247
Eurotium spp. 246
Xerophilic fungi 246
kidney damage 137
kippers 131
kiwi seed oil 112
Kreis test 33

labelling 217–220
additives 217–218
antioxidants 218
cholesterol 219
claims 218–220
durability 217–218
ingredients 217–218
PUFA 219
regulations 162
sequestrants 218
shelf life 218
storage 217–218
United States 219
vitamins 219–220
lactation cycle 182
'lactic' flavour 184
lactones 17, 18
lactoperoxidase
milk 186, 188
lard
induction period 59
oxidation 85, 86–88
shelf life 88
lauric fats
confectionery 246–247
lipolysis 24
law *see* legislation
LDL 138
'lean' fish 257
lecithin 236
structure 1
legislation 203–220
additives 209
antimicrobials 210–211, 214–215
antioxidants 210–211, 216
benzoate 211–213
cleaning 206
Codex Alimentarius 204–205, 207, 209
consumer protection 217
contaminants 209
dietary pattern 210
disinfection 206
EC 205
EC directives 205
EDTA 216
food handling 205–206
Food Safety Act 204
health hazards 206

legislation (cont.)
  history 203–204
  hydroxybenzoate 211–213
  hygiene 205–207
  hygiene regulation 206–207
  irradiation 208
  labelling 217–220
  migration through packaging 208
  packaging 208
  packaging gases 216
  positive list principle 209
  principles 203–205
  'quantum satis' 210
  sequestrants 211, 216
  sorbate 211–213
  sulphites 214–215
  sulphur dioxide 214–215
  trace metals 209
light
  packaging 74, 75
light-induced oxidation
  milk 185
linoleate, methyl
  autoxidation 11
linoleate, propyl
  accelerated oxidation 65
linoleic acid 7
linolenate, methyl
  decomposition 19
linolenic acid 7
linseed oil 130
lipase
  fungal 153
  wheat bran 143–146
  wheat germ 143
  assay 156–157
lipase assay 225
  non-dairy creams 224–225
lipase inhibitors 154, 181
lipases
  assay 24
  cereals (not wheat) 146–147
  milk 181–182
  oat 146
  rice 146–147
lipid composition
  meat tissues 193–194
lipofuchsin 136
lipolysis 7
  assay in dairy products 180
  butter 179, 184
  butter fat 183
  buttermilk 183–184
  coconut oil 246–247
  confectionery 243–244
  dairy products 179–184
  induced 181–182
  measurement 23–26

mechanism 17–18
milk 180–182
milk powder 183–184
palm kernel oil 246–247
pH-stat assay 183
post-manufacture 182–184
prevention 25
units 180
lipolysis assay
  indoxyl acetate 180, 183, 184
lipolysis in creams 223
lipolysis in milk
  agitation and 181–182
  spontaneous 181–182
lipolytic rancidity
  flavour thresholds in dairy products 184
lipoprotein lipase
  milk 181
lipoxidase see lipoxygenase
lipoxygenase 7, 9
  animal 262–265
  assay 107
  barley 152
  cereals 34, 143, 148–152
  co-oxidation 148
  co-substrate 148
  distribution within grain 148
  fish 262–265
  inhibitors 264
  milling and 149
  moisture 154
  nuts 248
  oat 151
  soybean 263
  specificity 263–264
  wheat 149
  wheat germ 150
  wheat, other than 151–152
lipoxygenase assay 34, 158–159
lipoxygenase reaction 14
liquid chromatography see HPLC
liver damage 134, 137
low density lipoprotein 138
LOX see lipoxygenase
luminol 35, 113–114

mackerel 129, 131
  non-enzymic initiation in 260
Maillard reaction 199
  confectionery 231, 238
  milk powder 245
Maillard reaction products
  antioxidants 227
malonaldehyde 124, 132
  cooking 132
  low density lipoprotein and 138
  mutagenicity 137

malondialdehyde 31
maltodextrin 227
management
  HACCP and 172, 176
manufacture
  critical control points 169
margarine
  fluorescence assay 117
MDA *see* malonaldehyde
measurement *see* assay
meat 191–201
  antioxidants in 199–200
  ascorbyl palmitate in 200
  bacon 196
  'bloom' 196
  comminution 192, 195
  compartmentation in 192
  cooking 132, 199
  cured 196
  enzymes in 192–193
  fat oxidation in 196
  fishy taints 199
  flavour 131
  freezing 196–197
  ham 196
  hydrolytic rancidity in 192–193
  microbial degradation 192–193
  myoglobin 194–196
  nitirites 198
  oxidative rancidity in 193–201
  oxygen in 194
  packaging 194, 196
  phospholipids 194
  pro-oxidants 201
  salt 198
  sausages 196
  shelf life 196–197
  species differences 193
  stale 196
  structure of 191–192
  sulphites 200
  sulphur dioxide 200
  surface oxygen 195
  tocopherols 199, 200
  'warmed over flavour' 199, 200
meat flavour 198–199
  phospholipids 199
meat oxidation
  temperature 196–197
meat pH 192
meat pigments 194–196
mechanism
  antioxidants 87
  autoxidation 8–12
  lipolysis 17–18
  photo-oxidation 12–14
melting points
  animal fats 197

membranes
  peroxides 134
metal levels
  oils and fats 161
metallic taste 243
metals
  cereals 152
methyl esters
  assay of 105
methylene group 9
met-myoglobin 194–195, 200
  initiation 259
MFGM *see* milk fat globule membrane
Michaelis constant 261
microbial degradation 192–193
  antioxidants 90
  dairy products 182–184
  water 237–241
microbial growth table
  water activity and 240–241
microsomal enzyme
  lipid oxidation 260–261
microwave cooking 132
migration
  packaging 208
milk
  ADV 186
  ascorbic acid in 187
  blood components in 182
  condensed 245
  copper 186, 188
  dissolved oxygen 186–187
  fat globule membrane 181–182
  fat globules 181–182
  fluorescence assay 117
  free fatty acids 244
  'high-linoleic acid' milk 187
  homogenisation 181, 187
  light-induced oxidation 185
  lipolysis 180–182
  oxidative rancidity in 185–187
  packaging 186
  processing and oxidation 188
  seasonal factors 182
  skimming efficiency 179
  soapy taste 179
  spontaneous oxidation 186
  storage 186, 187
  taste 186
  tocopherols in 187
  trace metals 187
  transportation 181
  triglycerides 179
  UHT 186, 187
milk fat globule membrane 226
milk lipases 181–182
milk lipolysis
  assay 180

milk powder
  chemiluminescence 61, 116
  GC-MS assay 124
  HPLC 113
  lipolysis 25, 183–184
  Maillard reaction 245
  oxidation 188
milk products *see* dairy products
milling
  lipoxygenase 149
  wheat lipase and 145
mincing *see* comminution
moisture
  assay 24
  cereal storage 154
  cereals 153
  lipoxygenase 154
  wheat lipase and 144
monitoring
  HACCP 174
muscle
  heart 135
mutagenicity 137–138
  malonaldehyde 137
myocarditis 136
myoglobin 194–196, 200, 258–259
myoglobin
  nitrosylated 196
myopathy 135

NAD 260–261
NADPH 260–261
nasogastric feeds 135
natural antioxidants 188, 235
  cereals 152
necrosis 135
Nickersen-Likens extraction *see* SDE
NIRD 225
nitrites in meat 198
nitrogen
  packaging 85
  protection from oxidation 79, 80
nitrogen storage
  powdered fats 227
nitrosamines 198
  ascorbic acid 198
nitroso-myoglobin 196
nitrous oxide 216
NMR 117
non-dairy creams 222–227
  assay 224–225
non-enzymic browning 232
non-enzymic initiation 258–260
non-enzymic oxidation
  cereals 152–153
nut oils
  fatty acid composition 250
nutrition 128–140
Nutrition Labelling Directive 219

nutritional quality 129
nuts 25
  autoxidation 252
  confectionery 248–253
  copper and iron 249
  flavour on storage 251–252
  oils 233
  peroxide value 249
  risk assessment 170

oats
  biscuits 162
  green shelling 162
  lipase 146
  lipoxygenase 151
  stabilisation 162
oct-1-ene-3-one 243
octa-1,5-dienone 243
octanoate
  assay 33
offal 129
off-flavours *see also* flavour
  alcohols 17
  aldehydes 16
  lipolysis 18
Oil Stability Index *see* OSI
oils
  addition of antioxidants 99
  bulk 76
  coldpressed 130
  nuts 233
oils and fats 161
  animals 3, 4
  blending 165
  bulk delivery 165
  buying policy 164
  confectionery 232–237
  metal levels 161
  stock control 164
  vegetable 5, 6
  world production 2
oleate, methyl
  autoxidation 10
oleic acid 7
oleo resins 163
organic free radicals
  in foods 106
orgoanoleptic assessment *see* taste
OSI
  assay 49–50
osmotic pressure 240–241
OV *see* oxodiene value
'own label' policy 165
oxidation
  and food processing 132–133
  babymilk 188
  butter 188
  carageenans and 188
  cereals 141–159

cream 188
  during transport 78–80
  functional properties 143
  in cooking 132
  low density lipoprotein and 138
  microsomal enzymes 260–261
  milk powder 188
  milk processing and 188
  spontaneous, in milk 186
oxidation in wheat
  assay 150–151
oxidation mechanism 1–18
  fish 257–265
  oxygen 258–259
oxidation products
  estimation 56
  in diet 129–131
  UV absorption 56
oxidation resistance assays 37–50
oxidative rancidity
  cereals 147–153
  dairy products 187–188
  meat 193–201
  milk 185–187
  salt 198
oxidised fat
  carcinogenicity 137–138
  epidemiology 137–138
  mutagenicity 137–138
'oxidised fat syndrome' 135
oxidised flavour 167
  milk 186
oxidograph 43
oxodiene value 61
oxygen
  assay 105
  bread-making 149
  dissolved 76
  dough 149
  meat 194–195
  mechanism of oxidation 258–259
  milk 186–187
  singlet 12, 13, 236
  solubility in oil 76–80
  triplet 12, 13
oxygen electrode 157–158, 159
ozone 201

packaging 74–76, 163
  antioxidants 75
  biscuits 173
  carbon dioxide 196
  controlled atmosphere 200–201, 269
  environment 166
  fish 268–269
  HACCP 173
  ink 173
  inks 163
  legislation 208

light 74, 75
meat 194
migration of substances 208
milk 186
myoglobin and 195, 196
nitrogen atmosphere 85
paper 173
photo-oxidation 173
polypropylene 163
powdered fats 228
snackfoods 75
trace metals 163
vacuum 94, 200–201, 216, 269
packaging gases
  legislation 216
PAH *see* polycyclic aromatic hydrocarbons
palm kernel oil
  ketonic rancidity 246–247
  lipolysis 246–247
palm oil
  lipolysis 25
  specification 235
  storage in drums 73
  transport 78–80
paper
  packaging 173
particle size
  flour 155
  wheat lipase 145
pasta 148, 152
pasteurisation 183
  creams 222
pasteurised milk 185–187
pathology
  clotting 138
  peroxides 134–136
peanut oil 112
pecan nuts
  tocopherol 249
pentane 115, 116
  assay 134
pentane value 57
permitted levels *see also* legislation 214–215
  antioxidant 101
peroxide assay 113, 114
  geometric isomers 111
  HPLC 109
peroxide formation
  chocolate 115
peroxide value 27–29, 58, 60, 234
  assay in dairy products 185
  correlation with HPLC assay 110, 112
  dissolved oxygen 76–80
  fish 265
  misleading 87
  nuts 249
  spectrophotometric 108

peroxide value (*cont.*)
  storage 70
  wheat 149
peroxides
  age 138
  atherosclerosis and 138
  biochemical effects 134
  breakdown 61
  clotting 138
  decomposition 57
  derivativization 114
  membranes 134
  pathology 134–136
  reduction 110
  spectrum 34
peroxy radical 8
pH
  meat 192
phenols
  antioxidants 19
phosphate in powdered fats 227
phosphatidylethanolamine
  structure 1
phosphatidylserine
  structure 1
phosphoglycerides
  structure 1
phospholipid peroxides 113
phospholipids 199
  cocoa butter 242
  meat 194
phospholipids
  milk fat globule membrane 226
phosphoric acid 89
photo-oxidation 188, 236, 258–259
  inhibitors 13
  mechanism 12–14
  packaging 173
  quenchers 14
  rates 234
  sensitisers 12
  type I 12
  type II 12
photosensitisers 12
  fish 258
pH-stat assay
  lipolysis 183
'pig' 72
pigments
  meat 194–196
pipelines 72
plant design
  HACCP 172
polar material 133
polarography 35
polycyclic aromatic hydrocarbons 132, 137
polymeric antioxidants 102

polymeric products 129
polypropylene
  biscuit packaging 163
polysaccharides
  confectionery 231
polyunsaturated fatty acids
  animal fats 131
  in diet 131
  in fish 131
  in foods 129–130
  in vegetable oils 130
positive list principle 209
post-manufacture lipolysis 182–184
potato chips 36
  antioxidants 95, 97
poultry
  irradiation 133
powdered fats 227
  antioxidants 228
  caseinate 227
  encapsulation 228
  hydrogenated oils 228
  maltodextrin 227
  packaging 228
  phosphate 227
  storage 227
process tanks 71
process vessels 69, 164
processing 153
  and oxidation 132–133
  cereals 152
  dissolved oxygen 76–80
  milk and oxidation 188
  oils and fats 164–165
  stabilisation 155
processing factors 76–82
product
  HACCP 172–173
product acceptability
  biscuits 167
  confectionery 167
product development 169
production
  HACCP and 175
promotion 218–220
pro-oxidants 59
  meat 201
propagation 8–12, 18, 57, 87
propyl gallate 20
  properties 93, 95
proteins
  confectionery 231–232
  oxidation inhibitors 133
psychrotrophs 183
PUFA *see* polyunsaturated fatty acids
pumping
  free fatty acids in milk 181
pumps 72

purge and trap 123
purification
 hydroxyacids 111
PV *see* peroxide value
pyrolysis
 carcinogenicity 137
pyrolysis of fats 129

quality
 water activity and 238
quality assurance 169
quality systems 207
'quantum satis' 210
quenchers
 photo-oxidation 14
quenching 236

radicals 8, 19
 allylic 9
 assay 106
rancid
 taste 135
rancidity
 butter 179
 cereals 141-159
 confectionery 171, 230-254
 control of 267-269
 creams 222-228
 dairy products 179-188
 definition 2, 128
 desserts 222-228
 hazards associated with 167-168
 'healthy eating' and 174-175
 history 160-161
 importance 160
 ingredients 161
 ketonic 22, 168, 246-247
 labelling 217-218
 legislation and health hazards 206
 measurement 22-53
 meat 191-201
 meat flavour 198-199
 nutrition 128-140
 nuts 248-253
 packaging and 163
 prevention by antioxidants 90
 process vessels 164
 processing 164-165
 sensory evaluation 55
 wheat germ 144
Rancimat 65, 236
Rancimat assay 46-49, 59, 176
Rancimat assay
 apparatus 47-48
rapeseed oil 138
RDA
 vitamins 219-220

reaction rate
 factors affecting 68
Recommended Daily Amount
 vitamins 219-220
refining
 dissolved oxygen 76-80
 temperature 80
refractive index
 assay 34
regression analysis
 assay correlation 63
regulations
 additives 162
 antioxidants 84, 100-101
 labelling 162, 166
resistance
 to oxidation 69
reversion 57
 flavour 167, 225
rice
 cooking 147
 lipase 146-147
rice bran 147
risk assessment 169-172
 environment 172
 flour 172
 season 172
roast meat
 flavour 199
ronoxan D20 200
rosemary 154
rosemary oil 113
rubbery flavour 236

safety
 antioxidants 86
safety assessment
 additives 210
salami 113
 antioxidants 124
 fluorescent products 124
sale
 HACCP 174
salmon
 oxidation 123
Salmonella 133
salt
 oxidation in meat 198
sausages 193, 196, 198
 'white spot' in 200
Schaal Oven Test 38, 58, 98
 history 160-161, 177-178
Schiff base
 assay 124
Scutellum 148
SDE 124
season
 risk assessment 172

seasonal factors
  milk   182
secondary oxidation products
  assay   117–124
secondary products   14–17, 19
  alcohols   16
  GC analysis   115, 116
  hydrocarbons   17
  table   104
seeds
  confectionery   248–253
selenium   134
sensitisers   12
sensory evaluation   55
sequestering agents   89, 91, 95, 161
  dairy products   188
  labelling   218
  legislation   211, 216
shallow frying   *see* frying
shelf life   54, 123, 165–166, 176
  cereals   122
  creams   129, 226
  fish   265
  labelling   218
  lard   88
  meat   196–197
  prediction   65
shellfish   129
short-life creams   222–226
simultaneous distillation extraction   124
singlet oxygen   12, 236
sitosterol   129
skimming efficiency
  milk   179
smoked food   137
snackfoods
  packaging   75
soapy flavour   23, 153, 160, 161, 162, 223
  coconut   161
  milk   179
solid creams   226–227
  shelf life   226
sorbate
  legislation   211–213
sorbate mould inhibitor   228
soup
  pentane production   115, 116
soybean flour
  co-oxidation   111, 149–150
soybean lipoxygenase   263
soybean oil
  induction period   60
'Spanish oil scandal'   138
sparging   78
  multi-stage   79
spectrophotometric assay
  hydroperoxides   106–108

peroxide value   108
spherosomes   152
spices
  use of   163
spin adducts   106
spontaneous lipolysis
  milk   181–182
spontaneous oxidation   186
spray-drying
  fats   227
stabilisation of oats   162
stale flavour
  milk   186
static headspace   121–123
steam coils   70, 71
steam heating   71, 162
steatitis   136
steel
  vessels   69, 70
sterilisation
  Alpura process   225
  creams   222
stock control   164
stomach cancer   137
storage
  beer   152
  cereals   154
  drums   72–74
  fish   260, 268
  flavour   75
  gallates   96
  HACCP   174
  'ink' formation   96
  labelling   217–218
  meat   196–197
  milk   186, 187
  powdered fats   227
  PV   70
  TBHQ   97
  temperature   80
  tocopherol   70, 153
  wheat lipase and   146
storage tanks   69
sulphites   200
  legislation   214–215
sulphur dioxide   200
  legislation   214–215
sunshine
  heating by   71
superoxide   187, 258–259
superoxide dismutase   134
  milk   187
surface area
  wheat lipase and   146
Swedish Food Institute   228
swift test   40, 45, 58, 65
Sylvester test   39
synergism
  antioxidants   89

synergy
  antioxidants   20, 269

tanks
  coatings   70
  heating   71
  storage   69
taste   135
  fishy   242–243
  HACCP   176
  hazelnuts, on storage   251
  metallic   243
  milk   186
  soapy, in milk   179
taste panels   22, 38
TBA test   31–33, 118
  fish   265
TBA value   132
  wheat   149
TBARS   118
TBHQ   20
  fish   269
  packaging   75
  properties   93, 96
temperature
  in refining   80
  oxidation in meat   196–197
  storage   80
termination reactions   8–12, 57
tetrabutyl hydroquinone   see TBHQ
Tetrapak   225
TG-OOH   see triglyceride hydroperoxides
thermogravimetric assay   61
thiobarbituric acid   see TBA
tocols
  cereals   152
tocopherols   20, 91–92, 134, 150, 158
  meat   199, 200
  milk   187
  nitrosamines   198
  storage   70, 153
  vegetable oils   235
  wheat   153
Totox value   31, 234
toxic oil syndrome   138
toxic products   128
toxicity
  acute   134
  atherosclerosis   138
  cardiomyopathy   136
  chronic   135
  diarrhoea   135
  diet   137
  growth rate   135
  haemolytic anaemia   136
  hepatomegaly   135
  kidney   137

liver damage   137
  myopathy   135
  polar fraction   137
  steatitis   136
  vitamin deficiency   136
toxicology   134–138
  long term   136–137
trace metals   161
  in body   134
  legislation   209
  milk   187
  packaging   163
transport   78–80
  milk   181
  palm oil   78–80
triglyceride hydroperoxides
  assay   109–114
triglycerides
  milk   179
  structure   1
trimethylamine   199
trioctanoin   25
triplet oxygen   12

UHT milk   186, 187
United States
  labelling   219
unsaturation
  assay   60
urine
  TBA assay   134
UV
  deactivators   20
UV absorption
  oxidation products   56
UV assay   63
  fish   266

vacuum packaging   194, 200–201
  fish   269
vanillin   154
vegetable fats   5, 6
  induction period   41
  polyunsaturated fatty acids in   130
  tocopherol content   91
verification
  HACCP   176
'vertical' directives   205, 207
vitamin C   see ascorbic acid
vitamin deficiency   136
vitamin E   see tocopherol
vitamins
  labelling   219–220
  RDA   219–220

'warmed over flavour'   199
water
  assay   24
  microbial degradation   237–241

water activity
  autoxidation 238
  browning 239
  flavour 237–241
  microbial growth table 240–241
water glaze 268
wheat
  carotenoids 150, 152
  durum 148, 152
  lipase 143–146
  lipoxygenase 149
  peroxide value 149
  TBA value 149
  tocopherol 153
wheat bran
  rancidity 144
wheat germ
  lipoxygenase 150
'wheat germ lipase' 143

  assay 145, 156–157
  milling and 145
  moisture 144
  particle size and 145
  storage and 146
  surface area 146
wheat products
  co-oxidation 149–150
'white spot' 200
WHO 204
wholegrain 141
wholemeal flour 150
Wolff titration 25
World Health Organisation see WHO

xanthine oxidase
  milk 186, 188
xerophilic fungi 246